Mathematical Programming Via Augmented Lagrangians

An Introduction with Computer Programs

APPLIED MATHEMATICS AND COMPUTATION

A Series of Graduate Textbooks, Monographs, Reference Works

Series Editor: ROBERT KALABA, University of Southern California

No. 1 MELVIN R. SCOTT
Invariant Imbedding and its Applications to Ordinary Differential Equations: An Introduction, 1973

No. 2 JOHN CASTI and ROBERT KALABA
Imbedding Methods in Applied Mathematics, 1973

No. 3 DONALD GREENSPAN
Discrete Models, 1973

No. 4 HARRIET H. KAGIWADA
System Identification: Methods and Applications, 1974

No. 5 V. K. MURTHY
The General Point Process: Applications to Bioscience, Medical Research, and Engineering, 1974

No. 6 HARRIET H. KAGIWADA and ROBERT KALABA
Integral Equations Via Imbedding Methods, 1974

No. 7 JULIUS T. TOU and RAFAEL C. GONZALEZ
Pattern Recognition Principles, 1974

No. 8 HARRIET H. KAGIWADA, ROBERT KALABA, and SUEO UENO
Multiple Scattering Processes: Inverse and Direct, 1975

No. 9 DONALD A. PIERRE and MICHAEL J. LOWE
Mathematical Programming Via Augmented Lagrangians: An Introduction with Computer Programs, 1975

Other Numbers in preparation

Mathematical Programming Via Augmented Lagrangians

An Introduction with Computer Programs

DONALD A. PIERRE
Systems Program, Electronics Research Laboratory
Montana State University
Bozeman

MICHAEL J. LOWE
Teledyne-Brown Engineering
Huntsville, Alabama, and
Department of Industrial and Systems Engineering
University of Alabama, Huntsville

 1975
Addison-Wesley Publishing Company
Advanced Book Program
Reading, Massachusetts

London · Amsterdam · Don Mills, Ontario · Sydney · Tokyo

Library of Congress Cataloging in Publication Data

Pierre, Donald A
  Mathematical programming via augmented lagrangians.

  (Applied mathematics and computation; no. 9)
  Bibliography: p.
  Includes indexes.
  1. Nonlinear programming. 2. Lagrangian functions.
I. Lowe, Michael J., joint author. II. Title.
T57.8.P53   519.7'6   75-22498
ISBN 0-201-05796-4
ISBN 0-201-05797-2 pbk.

Reproduced by Addison-Wesley Publishing Company, Inc., Advanced Book Program, Reading, Massachusetts, from camera-ready copy prepared by the authors.

Copyright © 1975 by Addison-Wesley Publishing Co., Inc.
Published simultaneously in Canada.

All rights reserved. No part of this publication may be reproduced, stored in retrieval system, or transmitted, in any form or by any means, electronic, mechanical, photocopying, recording, or otherwise, without the prior written permission of the publisher, Addison-Wesley Publishing Company, Inc., Advanced Book Program, Reading, Massachusetts 01867, U.S.A.

Manufactured in the United States of America

ABCDEFGHIJ-MA-798765

To

Mary   Patty

Mike   Krista

Louise   and

John   Larry

CONTENTS

SERIES EDITOR'S FOREWORD xiii

PREFACE xv

1. PROBLEM STATEMENT AND PRELIMINARY
   CONCEPTS 1

   1.1 Introduction 1

   1.2 The NLP in a Standard Form 4

   1.3 Representative Functions of
       One Variable 7

   1.4 Representative Functions of
       n Variables 14

   1.5 The Value of Convexity 21

   1.6 Sequential Unconstrained
       Extremization 28

   1.7 Brief History of Multiplier
       Methods 37

   1.8 Interlude 41

       Problems 43

2. CHARACTERIZATION OF CONSTRAINED
   MAXIMA: CONDITIONS OF OPTIMALITY . . . 49

   2.1 Introduction 49

		Page
2.2	Assumptions and Definitions	50
2.3	Taylor Series in n Variables	54
2.4	Conditions for Unconstrained Maxima and Minima	56
2.5	Case of One Equality Constraint	59
2.6	Case of Several Inequality Constraints	67
2.7	Abnormal Case	73
2.8	Preliminary Conditions for the General Case	78
2.9	Farkas Lemma	83
2.10	Existence of Generalized Lagrange Multipliers: Kuhn-Tucker Conditions	87
2.11	Sufficient Conditions for Constrained Maxima	93
2.12	Sufficient Conditions and the Linear Problem	100
2.13	Summary	102
	Problems	104
3.	SENSITIVITY	110
3.1	Introduction	110
3.2	Microscopic Sensitivity	111
3.3	Macroscopic Sensitivity	115
3.4	Sensitivity and Lagrange Multipliers	117

CONTENTS   ix

			Page
	3.5	Summary	124
		Problems	125
4.	AN AUGMENTED LAGRANGIAN AND THE STRUCTURE FOR A MULTIPLIER ALGORITHM		127
	4.1	Introduction	127
	4.2	An Augmented Lagrangian	128
	4.3	Properties of Unconstrained Maxima of L_a	131
	4.4	An Overview of a Multiplier Algorithm	144
	4.5	Finite Convergence for a General Linear Case	152
	4.6	Local Duality	154
	4.7	Multiplier Update Rule via Local Duality	165
	4.8	Convergence Rates as a Function of Weights	168
	4.9	Summary	170
		Problems	171
5.	IMPLEMENTATION OF THE ALGORITHM		175
	5.1	Introduction	175
	5.2	User Supplied Subroutines	179
	5.3	The General Routine LPNLP	180
	5.4	The Unidirectional Search Problem	199

			Page
	5.5	Conclusion	214
		Problems	216
6.	TEST PROBLEMS		218
	6.1	Introduction	218
	6.2	Unconstrained Test Problems and Results	219

   T1 Banana Function 219
   T2 Wood's Function 220
   T3 Powell's Function 220
   T4 Quartic 221
   T5 Biggs' Function (N = 3) 221
   T6 Biggs' Function (N = 5) 222
   Results for the Unconstrained Problems 223

  6.3 Constrained Test Problems and Results 228

   T7 Around the World 228
   T8 A Linear Problem 232
   T9 Seven Variable Problem 233
   T10 A Problem Suggested by Fiacco and McCormick 234
   T11 A Problem Suggested by Rosen and Susuki 234
   T12 A Problem Suggested by Beale 235
   T13 A Problem Suggested by Powell 236
   T14 A Problem Suggested by Wong 237
   T15 Another Suggested by Wong . . . 238
   T16 A Third Problem Suggested by Wong 239
   Results for Constrained Problems 242

  6.4 A Problem Not Satisfying Sufficient Conditions 245

CONTENTS   xi

                                                                    Page

    6.5 Summary 248

          Problems 253

7. APPLICATION PROBLEMS 259

    7.1 Introduction 259

    7.2 A Geometric Problem 262

    7.3 Deterministic Nonlinear Programming
          Equivalent for a Stochastic
          Linear Programming Problem --
          Feed Rationing 267

    7.4 A Circuits Problem 279

    7.5 Summary 289

          Problems 290

APPENDIX A. Programming Instructions for
            Problem Implementation
            Using LPNLP 295

    A.1 General Notation 295

    A.2 User Supplied Routines 298

    A.3 Information and Format of the
          Data Cards 302

    A.4 Control Cards 316

    A.5 Sample Case: Test Problem T9
          of Chapter 6 319

APPENDIX B. FORTRAN Program for LPNLP
            and Subroutines 326

APPENDIX C. Modified Lagrangian Forms . . . 363

APPENDIX D. Matrix Operations and
            Positive-Definite Tests . . . 368

		Page
APPENDIX E.	Unconstrained Search and Self Scaling	383
E.1	Introduction	383
E.2	Unconstrained Search	384
E.3	Steepest Ascent	386
E.4	Newton Search	388
E.5	Conjugate Gradient Methods	389
E.6	Method of Fletcher and Reeves	392
E.7	Quasi-Newton Methods	394
E.8	The DFP Method	395
E.9	Self-Scaling Variable-Metric Algorithm	399
E.10	Comment	407
REFERENCES		410
AUTHOR INDEX		426
SUBJECT INDEX		429

SERIES EDITOR'S FOREWORD

Execution times of modern digital computers are measured in nanoseconds. They can solve hundreds of simultaneous ordinary differential equations with speed and accuracy. But what does this immense capability imply with regard to solving the scientific, engineering, economic, and social problems confronting mankind? Clearly, much effort has to be expended in finding answers to that question.

In some fields, it is not yet possible to write mathematical equations which accurately describe processes of interest. Here, the computer may be used simply to simulate a process and, perhaps, to observe the efficacy of different control processes. In others, a mathematical description may be available, but the equations are frequently difficult to solve numerically. In such cases, the difficulties may be faced squarely and possibly overcome; alternatively, formulations may be sought which are more compatible with the inherent capabilities of computers. Mathematics itself nourishes and is nourished by such developments.

Each order of magnitude increase in speed and memory size of computers requires a reexamination of computational techniques and an assessment of the new problems which may be brought within the realm of solution. Volumes in this series will provide indications of current thinking regarding problem formulations, mathematical analysis, and computational treatment.

ROBERT KALABA

PREFACE

Solution techniques for the <u>nonlinear programming problem</u> (NLP) play a central role in mathematical programming and real-world problem solving and decision making. The NLP arises from many apparently unrelated areas and takes on many forms as found in business, economics, engineering, government, mathematics, and the natural and physical sciences.

In the abstract form the NLP is that of optimizing (maximizing or minimizing) some entity while satisfying side conditions or constraints. The entity to be optimized may be profit, cost, production, efficiency, consumer utility, social tension, social benefit, energy, inventory, etc., to find an optimal solution or a state of optimal operation.

Constraints are often imposed that may limit the possible steps or actions that may be taken to achieve this optimum. The constraints may involve manpower, availability of space, raw materials or funds, machine capabilities, limitations imposed by time, governmental controls, behavior patterns, physical laws, etc.

The NLP presents a solid framework for problem formulation. When an NLP is explicitly stated, a solution can generally be obtained. Many situations arise, however, where only numerical data or qualitative information is available and where the problem functions cannot be determined exactly, in which case the precise calculation of optimal points is impossible, but may be estimated with proper analysis. Of course, the wise decision maker does not blindly accept the optimal point for the NLP as the best answer for his real-world problem. The optimal point is considered only as advisory, and the assumptions and the accuracy of the NLP are scrutinized before a decision is made. In many instances, however, NLP solutions have proven themselves in practice to the extent that optimal points are accepted and implemented almost without question.

PREFACE

The transformation of a real-world problem into an explicit NLP by mathematical modeling is largely an art, and depends heavily upon the ability of man to interpret important conceptual causes and effects in terms of meaningful parameters and to relate them in mathematical form. Many different types of problems can be formulated as NLP's. The task of solving an NLP once it is formulated is generally not a trivial one. Efficient, convenient, and robust methods of solution are actively sought.

The purpose of this book is threefold: 1) to treat a general nonlinear programming model theoretically while giving insight into the basic conditions of optimality upon which solution algorithms and methods of sensitivity can be based; 2) to introduce a powerful method of nonlinear programming that employs augmented Lagrangians, leading to a comprehensive algorithm and computer code for solving NLP's that may depend on many variables and constraints: and 3) to provide test problems and application problems that convey the breadth of nonlinear programming in addition to illustrating some of the art of converting real-world problems into NLP's.

Numerical techniques of nonlinear programming have been treated in the past mostly with emphasis on the development of computer codes for the constrained minimization problem. In this work we emphasize the constrained maximization problem not to be different, but with the feeling that anyone familiar with linear programming codes may relate more easily to this development of the nonlinear problem. In many of those situations where linearized models have been treated in lieu of a nonlinear model, the linear approximation may be discarded and the true nonlinear problem solved directly. The nonlinear code can also be used to solve the linear problem; in fact, the computer program of this text solves a broad class of linear programming problems in a finite number of steps. For those familiar with other methods of nonlinear programming, hopefully, new thoughts will be developed on the use of augmented Lagrangians in mathematical programming.

The text can be used at the junior level or above for a one-quarter three-credit course.

PREFACE

Familiarity with set notation, with vectors and matricies, and with derivatives of functions is assumed background. Over the past 12 years, Dr. Pierre has taught material similar to that contained in Chapters 1, 2, 3, 6, and 7 as part of a three-quarter course sequence entitled "Optimal Design of Systems". The material in Chapters 4 and 5 is relatively new: since 1971, the authors have collaborated in making contributions to the study of augmented Lagrangians.

Chapter 1 contains the statement of the NLP in a standard form. Representative functions that are encountered in problem formulations are considered. When these functions possess appropriate convex or concave properties, it is shown that the corresponding NLP solution has a single local optimal value. Background literature and related solution concepts are also reviewed in Chapter 1. Chapter 2 contains a systematic development of first-order and second-order optimality conditions that apply to a regular point of an NLP. In Chapter 3, major emphasis is placed on the role that Lagrange multipliers play in the sensitivity analysis of NLP solutions. Chapter 4 builds on the results of Chapters 2 and 3, developing the rationale for multiplier algorithms; the multiplier update rule for a particular algorithm is derived through concepts of local duality. In Chapter 5 and Appendix E, numerical techniques for implementing multiplier algorithms are considered. Examples are given throughout the text to illustrate conceptual points, and problems are listed at the end of each chapter. In Chapter 6, test problems are given as a way of comparing alternative NLP solution methods; in particular, the FORTRAN code of Appendix B is shown to be superior to several others. The comprehensive examples of Chapter 7 give insight into ways of converting application problems into NLP's, and Appendix A contains computer programming instructions for solving NLP's using the FORTRAN code of Appendix B. Upon request, the authors will supply information on how to obtain a card-deck source for the code.

We thank the following people: Dr. Paul E. Uhlrich and Dr. Byron J. Bennett for their support and encouragement in this work; Mrs. Jean Julian for her excellent typing of exacting material; Mr. Chad A. Groth for his ability to transform rough sketches into inked art work; Mr. Douglas Winterrowd for programming and testing several alternative computer codes; and Mr. Ray S. Babcock for providing quality listings of computer programs. We are indebted to colleagues and former students whose questions often resulted in the clarification of subject matter. We are grateful for the financial support of this work by the Engineering Experiment Station, the Computer Center, and the Electrical Engineering Department of Montana State University. Finally, for their patience and understanding, we dedicate this book to our families.

    Donald A. Pierre   and   Michael J. Lowe

Mathematical Programming
Via Augmented Lagrangians

An Introduction with Computer Programs

I
PROBLEM STATEMENT AND PRELIMINARY CONCEPTS

1.1 INTRODUCTION

The nonlinear programming problem (NLP) is the problem of extremizing (maximizing or minimizing) a function of n variables while requiring other functions of the same variables to satisfy either equality or inequality constraint relationships. Section 1.2 contains the mathematical description of the general NLP considered in most of the text -- the variables are assumed to be continuous, rather than discrete, and the functions involved should exhibit a specified degree of continuity. Sections 1.2 and 1.3 contain examples of such functions. When all the functions involved are linear, the NLP reduces to a linear programming problem. Now, it is a generally accepted viewpoint that nature is nonlinear. In terms of application, however, nonlinear programming to date has held second place to linear programming. The would-be paradox of the preceding two statements can be explained as follows:

Linear programming problems have an elegant structure which led Dantzig [D1] to the development of the simplex solution algorithm at the very beginning of work in that area.[†] In contrast, whereas the fundamental theoretical paper on nonlinear programming by Kuhn and Tucker [K6] appeared in 1951, the associated structure of the nonlinear problem did not appear to be as conducive to the development of solution algorithms. Indeed, it was not until 1960 that the first efficient nonlinear programming method (for the linear constraint case at that) was produced by Rosen [R7].

An allaying feature of linear programming is that it always leads to global optimal solutions for well-formulated linear problems. Only for special problem structures can nonlinear problems

---

[†] References are associated with letters and numbers in brackets and are listed at the end of the text.

1.1 PRELIMINARY CONCEPTS   3

be solved with such assurance; that is, depending on the nonlinear problem and on the starting values assigned to the variables, a given nonlinear programming algorithm may converge to any one of several different local optimal solutions. The development in Sections 1.3 and 1.4 provides one well-known class of nonlinear problem which, like the linear programming problem, exhibits one local optimum which is the global optimum.

When our problem is to maximize or minimize a function of n variables without regard to side conditions or constraints, we have an <u>unconstrained NLP</u>. The approaches that have been proposed to solve unconstrained NLP's could fill volumes. While research continues in this area, several good methods are available, some including aspects of self scaling. Appendix E contains descriptions of a few of these methods. Unconstrained search methods are often embedded in algorithms designed to solve the general NLP. Sequential unconstrained extremization (Section 1.6) is a classic example of this. By relegating unconstrained search algorithms to an appendix of this text, our intent is not to downgrade these methods -- it is only that the major thrust of this work is to present an effective overall method for solving constrained NLP's. The history of this method and of related methods is reviewed in Section 1.7.

1.2 THE NLP IN A STANDARD FORM

Let x denote the vector $(x_1, x_2, \ldots, x_n)$ of real variables $x_1, x_2, \ldots, x_n$. The vector x is an element of n-dimensional Euclidean space E^n (that is, $x \in E^n$), and particular values of x are called points in E^n. In an NLP, we have a function $f(x)$ which maps each x into a scalar value and which is to be extremized (maximized or minimized) with respect to x. The function $f(x)$ is called an objective function, return function, cost function, or performance measure; these names associated with $f(x)$ are used indiscriminately throughout the text. Since maximization and minimization are equivalent problems (the maximum of f and the minimum of -f occur at the same point or max $f = -\min(-f)$), only the former is emphasized here. The point x, however, may not be chosen arbitrarily because of constraints imposed upon it.

1.2 PRELIMINARY CONCEPTS

The constraints are generally classified as <u>equality</u> or <u>inequality</u> constraints and are defined in terms of x. Any x which satisfies all of the constraints is termed a <u>feasible point</u> and becomes a candidate for the optimal point. For a solvable problem there exists a nonempty set of feasible points. From this set of feasible points, a point x^*, at which f achieves its maximum value, is called an <u>optimal point</u>.

Stated mathematically, the NLP takes the following form: find a point x^* to

$$\text{maximize } f(x) \tag{1-1}$$

subject to the constraints

$$p_i(x) = a_i, \quad i = 1, 2, \ldots, m_1 < n \tag{1-2}$$

$$q_j(x) \leq b_j, \quad j = 1, 2, \ldots, m_2 \tag{1-3}$$

and

$$c_k \leq x_k \leq d_k, \quad k = 1, 2, \ldots, n \tag{1-4}$$

where the a_i's, b_j's, c_k's, and d_k's are real constants. The p_i's, q_j's, and f are called the <u>NLP functions</u> and can be linear or nonlinear.

Assumptions in regard to the NLP functions vary from place to place in the text, depending on the particular results to be obtained. Thus, for example, if appropriate convexity assumptions in Section 1.5 are satisfied, any constrained local maximum is shown to be global. Or, if the NLP functions have continuous first and second derivatives in the neighborhood of a constrained local maximum and if the gradient vectors[†] ∇p_i and ∇q_j are not identically zero at any point which satisfies $p_i = a_i$ and $q_j = b_j$, respectively, as is assumed in Section 2.11, then second-order sufficient conditions for constrained local maxima are relevant.

Problems 1.1 through 1.9 describe a few of the many possible situations that can be cast into the NLP format of (1-1) through (1-4). Constraints (1-4) permit us to impose explicit upper and lower bounds on each variable. These constraints are a

---

[†] Notation used is $\nabla p_i = \partial p_i/\partial x = (\partial p_i/\partial x_1, \partial p_i/\partial x_2, \ldots, \partial p_i/\partial x_n)$. In subsequent chapters, we also use column matrices to express vectors.

1.2 PRELIMINARY CONCEPTS

special case of constraints (1-3) because (1-4) is equivalent to

$$-x_k \leq -c_k$$
$$k = 1, 2, \ldots, n \qquad (1-5)$$
$$x_k \leq d_k$$

In the theory that follows, therefore, we condense the presentation by considering only the general constraints (1-2) and (1-3). In the computer programs, however, the special properties of (1-4) are used to simplify the preparation of problems for computer solution.

The problem as stated seeks the constrained absolute maximum of f in (1-1). In general practice, however, constrained local maxima of f are found first; the absolute maximum of f is then found by comparing all constrained local maxima.

1.3 REPRESENTATIVE FUNCTIONS OF ONE VARIABLE

A given NLP is characterized in general both by linear and nonlinear functions. Many nonlinear functions are useful in modeling NLP's -- the several that are given in this section are typical and exemplify some of the general properties of nonlinear functions. Figures 1-1 through 1-8 display the functions. In the following development we give a few of the many possible physical interpretations that can be assigned to the functions which could be incorporated either in an objective function (1-1) or in constraint relationships (1-2) and (1-3). In most applications, a given function will be found to characterize a given physical situation only in an approximate sense.

Consider Figure 1-1. The value y could represent, for example, the cost/hour of producing x units/hour of a given product. In this case, the constant c represents a fixed cost/hour incurred even if no units are produced; the constant b gives the linear rate at which cost/hour increases with the number of units/hour, and the constant $a > 0$ is associated with a quadratic cost term. If $a < 0$ in the above application, the function is not a realistic cost function for $x > x_m$ where $x_m = -b/2a$ is the value of x which yields the maximum value of y.

1.3 PRELIMINARY CONCEPTS

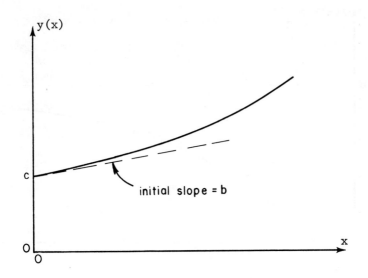

Figure 1-1 $y(x) = ax^2 + bx + c$ with $a > 0$ and $b > 0$.

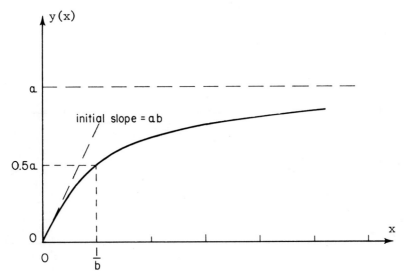

Figure 1-2 $y(x) = abx/(1 + bx)$ with $a > 0$ and $b > 0$.

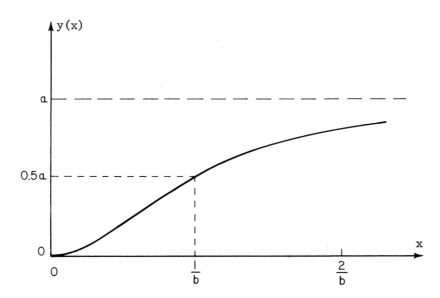

Figure 1-3 $y(x) = a(bx)^2/[1 + (bx)^2]$ with $a > 0$ and $b > 0$.

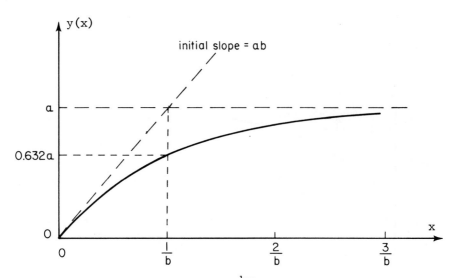

Figure 1-4 $y(x) = a(1 - e^{-bx})$ with $a > 0$ and $b > 0$.

1.3 PRELIMINARY CONCEPTS

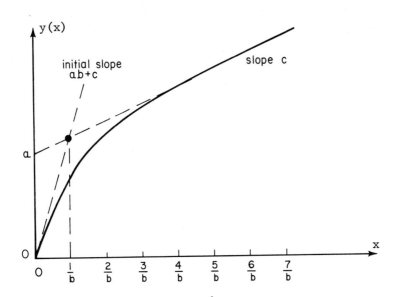

Figure 1-5 $y(x) = a(1 - e^{-bx}) + cx$ with a, b, and c > 0.

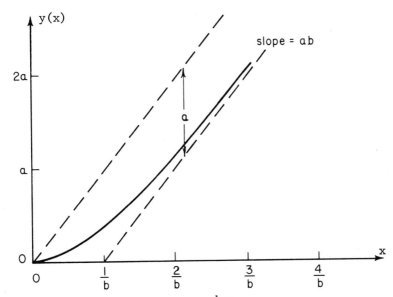

Figure 1-6 $y(x) = a(bx + e^{-bx} - 1)$ with a > 0 and b > 0.

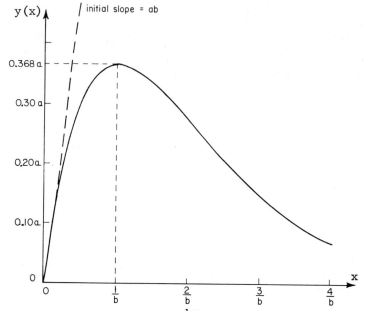

Figure 1-7 $y(x) = abxe^{-bx}$ with $a > 0$ and $b > 0$.

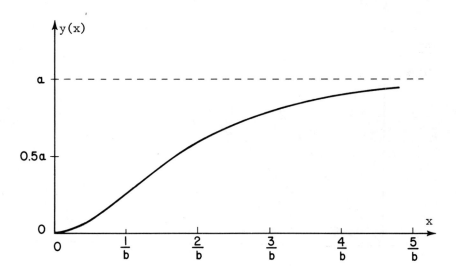

Figure 1-8 $y(x) = a(1 - e^{-bx} - bxe^{-bx})$ with $a > 0$ and $b > 0$.

1.3 PRELIMINARY CONCEPTS

Figures 1-2 and 1-4 characterize a saturation effect. For example, y in these cases could represent the amount of a specific product that could be produced by a given factory in an hour as a function of a particular raw material, represented by x, that the factory uses in an hour. Clearly there is an output limitation even for an infinite supply of the raw material. Figures 1-3 and 1-8 are even more characteristic of the above physical situation: in addition to the saturation effect, they give that there is a threshold of x below which the output produced is negligible.

Figure 1-5 could be representative of the following. Let y denote the cost of producing x units of a given product. To produce no units, it is given that there is zero cost, but to produce one or more units requires a set-up charge of approximately "a" dollars, in addition to c dollars per unit. To have Figure 1-5 conform to this situation, let b = 4 so that $y \approx a + cx$ for $x \geq 1$, and y = 0 for x = 0.

In Figure 1-6, the initial slope of 0 at x = 0 could be used to characterize the following. Suppose that company A produces a by-product which is normally viewed as scrap material: company A may be able to make some beneficial use of this scrap in other areas of its operation. Let x denote the amount of scrap material used and y denote its cost. When x is small, say $x < 0.5/b$, the cost of the scrap is low; but as x increases beyond that point where the material is readily produced as a by-product, company A may have to purchase the scrap material from other sources at a cost of ab dollars per unit of scrap material.

Each function shown in Figures 1-1 through 1-8 has continuous derivatives of all orders for $x \geq 0$. Also, each function shown, except that of Figure 1-7, is a <u>monotonic increasing</u> function of x for $x \geq 0$. Furthermore, each function shown has but one local maximum with respect to x for $0 \leq x \leq d$ where d is a given upper bound; each function therefore exhibits a <u>unimodal maximum</u> in the interval $0 \leq x \leq d$.

1.4 REPRESENTATIVE FUNCTIONS OF N VARIABLES

Functions of n variables assume many diverse forms. Here we review some of the general

1.4 PRELIMINARY CONCEPTS

properties of important classes of functions and consider a few examples of such functions -- many more examples are considered throughout the book.

One important class of functions is that in which each function of n variables can be represented by a sum of functions of one variable. The term <u>separable function</u> is used to describe such functions.

DEFINITION 1.1 [Separable Function]
A separable function f(x) of n variables is one which can be represented by a summation of the form

$$f(x) = \sum_{i=1}^{n} y_i(x_i) \qquad (1-6)$$

As an example of a separable function, consider the case in which a company produces n products. Let $y_i(x_i)$ denote the cost of producing x_i units of the ith product. Each $y_i(x_i)$ function might assume one of the forms given in the previous section. The total cost f(x) of producing the n products is a separable function $\sum_{i=1}^{n} y_i(x_i)$ under the assumed conditions. On the other hand, if the cost of producing x_i units of the ith product depends on the number x_j of the units of the jth product, in addition to depending on x_i, where $i \neq j$, then $f(x)$ may not be a separable function.

Another important class of functions is that in which each function exhibits only one local maximum (minimum). To be precise, the following definitions are in order.

DEFINITION 1.2 [Local Maximum over a Closed Set][†]
Let S be a closed set with $S \subset E^n$. Given a function $f(x)$ defined over S, a <u>local maximum</u> of $f(x)$ for $x \in S$ exists at $x^* \in S$ if there is a real number $\varepsilon > 0$ such that

$$f(x^*) \geq f(x^* + \Delta x) \qquad (1-7)$$

[†] Notation used is as follows: 1) $S \subset E^n$ gives that S is a subset of Euclidean space E^n; 2) $\{\Delta x \mid \cdots\}$ denotes a set of Δx's with the property that $\cdots$; and 3) $\|\Delta x\|$ denotes the Euclidean norm $[\sum_{i=1}^{n}(\Delta x_i)^2]^{\frac{1}{2}}$ of the vector Δx.

1.4 PRELIMINARY CONCEPTS

for all Δx in the set $\{\Delta x |\ 0 < ||\Delta x|| < \varepsilon$ and $(x^* + \Delta x) \in S\}$.

A <u>local minimum</u> of f over S is defined in the same way, but with the inequality in (1-7) reversed.

DEFINITION 1.3 [Unimodal over a Closed Set]
A function of n variables is <u>unimodal over a closed set</u> $S \subset E^n$ if there is but one local maximum (minimum) of $f(x)$ for $x \in S$. The one local maximum (minimum) is the global maximum (minimum) of $f(x)$ for $x \in S$.

For example, the nonseparable function $f(x) = x_1(1 - e^{-x_2})$ can be shown to be unimodal over the closed set S defined by $0 \leq x_1 \leq d_1$ and $0 \leq x_2 \leq d_2$ for any given positive values of d_1 and d_2.

Simplifications usually result if the closed set S of Definitions 1.2 and 1.3 is a convex set.

DEFINITION 1.4 [Convex Set]
Let $x(\rho) \triangleq \rho x^a + (1 - \rho) x^b$ where x^a and x^b are

contained in a closed set S.[†] The closed set S is said to be a <u>convex closed set</u> if and only if $x(\rho) \in S$ for all possible choices of x^a and $x^b \in S$ and for all ρ that satisfy $0 \leq \rho \leq 1$.

Figures 1-9a and 1-9b illustrate convex sets. A nonconvex set is shown in Figure 1-9c.

Before considering relationships between certain types of NLP constraint functions and convex sets, we must first review the concepts of convex and concave functions.

DEFINITION 1.5 [Convex (Concave) Functions]
Let $x(\rho) \triangleq \rho x^a + (1-\rho)x^b$. A function $y(x)$ is <u>convex</u> (<u>concave</u>) over a convex set S if $y(x(\rho))$ is less than (greater than) or equal to $\rho y(x^a) + (1-\rho)y(x^b)$ for all $\rho \in [0, 1]$ and for all $x^a, x^b \in S$.[‡]

[†] Here, with x^a and x^b, superscripts on x are used to label different points in E^n. This usage and that in which a variable is raised to a power are readily distinguishable from the context in which superscripts appear.

[‡] Here [0, 1] denotes the closed set of real values from 0 to 1.

1.4 PRELIMINARY CONCEPTS   19

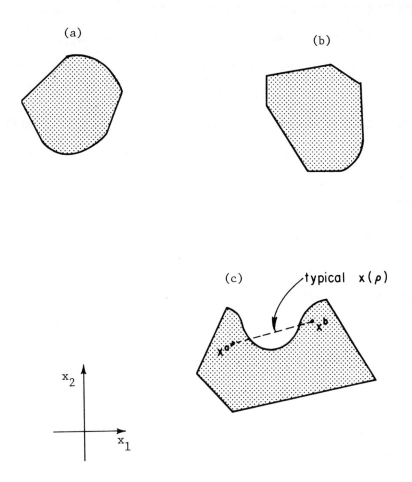

Figure 1-9 (a) Convex set in E^2. (b) Convex set in E^2.
(c) Nonconvex set in E^2.

The functions of Figures 1-1 and 1-6 are convex functions for $x \geq 0$; the functions of Figures 1-2, 1-4 and 1-5 are concave functions for $x \geq 0$; whereas the functions of Figures 1-3, 1-7 and 1-8 are neither convex nor concave for $x \geq 0$. Note that if $y(x)$ is convex, then $-y(x)$ is concave, and vice versa. Another important property of convex (concave) functions is the following.

THEOREM 1.1 [On Sums of Convex (Concave) Functions] The sum of convex (concave) functions is a convex (concave) function.

Proof. Let $f(x)$ equal $\sum_{i=1}^{k} g_i(x)$ where each $g_i(x)$ is concave on a convex set S. We have then, for $\rho \in [0, 1]$ and $x^a, x^b \in S$,

$$g_i(\rho x^a + (1-\rho) x^b) \geq \rho g_i(x^a) + (1-\rho) g_i(x^b) \quad (1-8)$$

from which it follows that

1.4 PRELIMINARY CONCEPTS

$$\sum_{i=1}^{k} g_i(\rho x^a + (1-\rho)x^b) \geq \rho \sum_{i=1}^{k} g_i(x^a)$$

$$+ (1-\rho) \sum_{i=1}^{k} g_i(x^b) \quad (1-9a)$$

or

$$f(\rho x^a + (1-\rho)x^b) \geq \rho f(x^a) + (1-\rho)f(x^b) \quad (1-9b)$$

for $\rho \in [0, 1]$ and $x^a, x^b \in S$. Thus, f is concave on S.

●

The proof for the case where f is the sum of convex functions is similar; only the inequalities in (1-8) and (1-9) need to be reversed.

1.5 THE VALUE OF CONVEXITY

Generally applicable nonlinear programming techniques isolate local extrema, a fact which leads to the following question: "Under what conditions can we guarantee that only one local maximum (minimum) exists and that this local maximum (minimum) is the absolute maximum (minimum)?" An answer to this is based on convex sets and on concave (convex) functions defined on convex sets. The purpose of this section is to give a well-known sufficient condition [K6], in terms of functions involved in the NLP, that guarantees that a local maximum (minimum) is also the global maximum (minimum). If the condition is not satisfied, there may yet be but one local maximum (minimum) -- computational methods are generally required to determine the number and location of constrained local extrema when appropriate convexity conditions are not satisfied.

Convex sets (Definition 1.4) are often specified in terms of a set of constraints that must be satisfied. We show this by the following theorem.

THEOREM 1.2 [Convex Set From Inequality Constraints]
Given a vector function $q(x)$ and a constant vector b, let S be the set of x's that satisfy $q(x) \leq b$; that is, $S \triangleq \{x| q(x) \leq b\}$. S is a convex set if the components of q are convex functions.

Proof. Given that $q(x)$ is a convex function,

1.5 PRELIMINARY CONCEPTS

$$q(x(\rho)) \triangleq q(\rho x^a + (1-\rho)x^b)$$
$$\leq \rho q(x^a) + (1-\rho)q(x^b) \qquad (1\text{-}10)$$

for $\rho \in [0, 1]$. Using any x^a and x^b contained in S, so that $q(x^a) \leq b$ and $q(x^b) \leq b$, the right-hand member of (1-10) can be replaced by $\rho b + (1-\rho)b$ to obtain

$$q(x(\rho)) \leq b \qquad (1\text{-}11)$$

and therefore $x(\rho) \in S$. Thus, S is convex.

●

The following theorem is a straightforward counterpart to Theorem 1.2 and is given without proof.

THEOREM 1.3 [Convex Set From Inequality Constraints] Given a vector function $q(x)$ and a constant vector b, let $S \triangleq \{x | q(x) \geq b\}$. S is a convex set if the components of q are concave functions.

We emphasize that Theorems 1.2 and 1.3 give sufficient conditions for S to be convex; the conditions are not necessary ones -- if $q(x)$ is not a convex function, which is contrary to the assumption in Theorem 1.2, the set $S = \{x \mid q(x) \leq b\}$ may yet be convex.

Given two or more constraints that must be satisfied, if each constraint defines a convex set, then the totality of the constraints defines a convex set. We formalize the previous statement as follows.

THEOREM 1.4 [Intersection of Convex Sets]
The intersection of convex sets is also a convex set.

Proof. Let S_1 and S_2 be convex sets and let S be the intersection of S_1 and S_2; that is, $S = S_1 \cap S_2$ which is assumed to be nonempty. Given any two points x^a and $x^b \in S$, then both x^a and x^b are contained in S_1 and S_2, as also is $x(\rho) = \rho x^a + (1 - \rho) x^b$, $0 \leq \rho \leq 1$, because S_1 and S_2 are convex. Thus, $x(\rho)$ is contained in S which is therefore convex.
●

Figure 1-10a illustrates the intersection of two convex sets. Figure 1-10b illustrates a case where the intersection of two nonconvex sets is

1.5 PRELIMINARY CONCEPTS 25

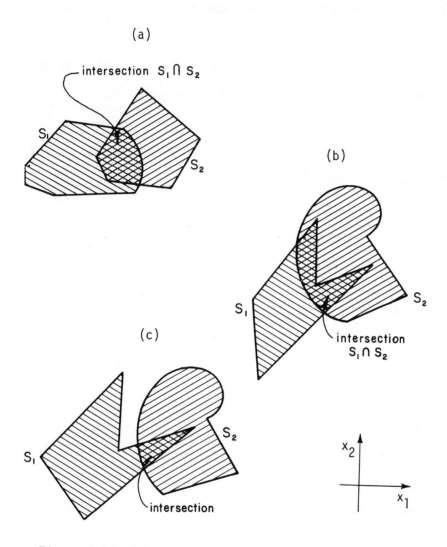

Figure 1-10 (a) Convex intersection of convex sets.
(b) Nonconvex intersection of nonconvex
sets. (c) Convex intersection of
particular nonconvex sets.

nonconvex; but note that Figure 1-10c illustrates a case where the intersection of two nonconvex sets is convex.

For equality constraints, the preceding theorems can be combined to give Theorem 1.5.

THEOREM 1.5 [Convex Set Defined by Linear Equalities] The set $S = \{x \mid p(x) = a\}$ is convex if $p(x)$ is linear.

Proof. Let $S_1 = \{x \mid p(x) \leq a\}$ and $S_2 = \{x \mid p(x) \geq a\}$. Theorem 1.2 gives that S_1 is a convex set because a linear $p(x)$ is a convex function. Theorem 1.3 gives that S_2 is a convex set because a linear $p(x)$ is also a concave function. Thus, S, which is the intersection of S_1 and S_2, is a convex set by Theorem 1.4. ●

We are now ready for the main theorem of convex programming.

THEOREM 1.6 [A Condition for a Global Maximum] Given any concave (convex) function $f(x)$ defined over a closed convex set S, the absolute maximum

1.5 PRELIMINARY CONCEPTS

(minimum) of f over S is the only local maximum (minimum) of f over S.

Proof. Assume the contrary -- let $f(x^*)$ be the absolute maximum over S, and assume that $f(x^0)$ is a local maximum with $f(x^0) < f(x^*)$. Because $f(x)$ is concave,

$$f(\rho x^* + (1-\rho)x^0) \geq \rho f(x^*) + (1-\rho)f(x^0), \quad 0 < \rho < 1 \quad (1\text{-}12)$$

Let $x(\rho) = \rho x^* + (1-\rho)x^0$, $0 < \rho < 1$. Because the set S is convex, all points $x(\rho)$ on the straight line segment between x^* and x^0 are in S. Also, with $f(x^0) < f(x^*)$, (1-12) can be replaced by

$$f(x(\rho)) > \rho f(x^0) + (1-\rho)f(x^0) = f(x^0) \quad (1\text{-}13)$$

Given any $\varepsilon > 0$, (1-13) indicates that a $\rho \neq 0$ exists for which the Euclidean norm $\|x(\rho) - x^0\|$ is less than ε and for which $f(x(\rho)) > f(x^0)$. Thus, $f(x^0)$ cannot be a local maximum.

●

1.6 SEQUENTIAL UNCONSTRAINED EXTREMIZATION

Basic to all NLP solution algorithms is a search in n-space for constrained extrema. Many constrained search techniques depend heavily upon the classical theory and methods for computing unconstrained extrema. Unconstrained search techniques often incorporate a sequence of unidirectional searches in n-space, and are based on an iterative scheme of the form†

$$x^{k+1} = x^k + \rho r^k \qquad (1\text{-}14)$$

where a parametric step ρ is taken in a vector direction r^k. The step ρ is selected to obtain the maximum value of f in the r^k direction. The search direction r^k is usually chosen on the basis of information concerning f and its partial derivatives evaluated both at the point x^k and at previous iteration points. The influence of constraints imposed on x must be taken into consideration as

---

† Here the superscript notation is used to denote a change in value of a vector or parameter from one iteration to another.

1.6 PRELIMINARY CONCEPTS

they alter the way in which the constrained search direction is chosen. This can be an intricate task if the constraints are nonlinear. According to Davies and Swann [D3], there are two general ways in which constraint information is incorporated in the establishment of search directions. They are as follows:

FUNCTION MODIFICATION FOLLOWED BY UNCONSTRAINED OPTIMIZATION. These methods seek to define a new function that has an unconstrained optimum at the same point as the optimum of the given constrained problem. Optimization of this new function will then define the required change in the search direction.

DIRECTION MODIFICATION WITHOUT ALTERING THE FUNCTION. Some of the methods in this category attempt to follow constraint boundaries while others try to rebound from them, so as to continue the search in the feasible region.

There are advantages and disadvantages to both types of methods which are explored in detail by

McCormick [M1] and Wolfe [W2]. No one method for constrained optimization is universally superior to all others and care must be exercised in choosing the appropriate one for a given situation.

The solution techniques developed in this work fall into the first category. The approach in these methods is that of transforming a given constrained maximization problem into a sequence of unconstrained maximization problems as outlined by Fiacco and McCormick [F6] for the minimization problem. This transformation is accomplished by augmenting the original cost function with weighted terms of the problem functions, to define a new objective function whose extrema are unconstrained in some domain of interest. By gradually removing the effect of the constraints in the new objective function by controlled parameter values, it is possible to generate a sequence of unconstrained problems that have solutions converging to a solution of the original constrained problem.

Thus, a typical sequential unconstrained method of maximizing the NLP is one associated with finding a sequence of problem solutions x^m such that

1.6 PRELIMINARY CONCEPTS

$$A(x^m, w^m) = \max_x A(x, w^m), \quad m = 1, 2, \ldots \quad (1\text{-}15)$$

where $A(x, w^m)$ is the new objective function formed by augmenting the objective f with weighted terms that depend on the constraints, and w^m is a controlled weighting factor. The functions of (1-15) are constructed so that as $m \to \infty$, a convergent sequence of unconstrained maximizers $\{x^m\}$ approaches a constrained maximum point x^* of the NLP.

This text focuses on a generalization of (1-15) in which w^m is replaced by a vector of weights and Lagrange multipliers. Several methods of this type have appeared during the past few years; depending on the author, these methods are called <u>multiplier methods</u>, <u>augmented Lagrangian methods</u>, or <u>dual methods</u>. The appropriateness of these names is made clear in the chapters that follow.

As an example of a particular method associated with (1-15), consider an augmented function $A(x, w^m)$ which incorporates the equality constraints by Courant's quadratic penalty function [C6], and the inequality constraints by Carroll's inverse penalty function [C2]. Define

$$A(x,w^m) \triangleq f(x) - w^m \sum_{i=1}^{m_1} (a_i - p_i(x))^2$$

$$- \frac{1}{w^m} \sum_{j=1}^{m_2} \frac{1}{(b_j - q_j(x))} \qquad (1\text{-}16)$$

To use this function we require a starting point x^0 which is in the strictly feasible region R^S associated with the inequality constraints, i.e.,

$$x^0 \in R^S \triangleq \{x \mid q_j(x) < b_j,$$

$$j = 1, 2, \ldots, m_2\} \qquad (1\text{-}17)$$

Because the function $A(x,w^m)$ is to be maximized, and because $A(x,w^m)$ assumes the value of $-\infty$ at the boundary of R^S, an unconstrained search applied to maximize $A(x,w^m)$ will generate only feasible points with respect to the inequality constraints (1-3), i.e., all points generated will remain in R^S.

The method generally proceeds as follows. Select a monotonic sequence $\{w^m\}$, where

1.6 PRELIMINARY CONCEPTS

$$\{w^m\} \triangleq \{w^m \mid w^m > 0,\ w^{m+1} > w^m,$$

$$\text{and}\quad w^m \to \infty \text{ as } m \to \infty\} \qquad (1\text{-}18)$$

and compute a maximum point x^m, where

$$A(x^m, w^m) = \max_{x \in R^s} A(x, w^m) \qquad (1\text{-}19)$$

for $m = 1, 2, \ldots$. The desirable result is that

$$\lim_{m \to \infty} x^m = x^* \qquad (1\text{-}20)$$

where x^* is a solution to the NLP. It is also well known [F6] that

$$\lim_{m \to \infty} w^m \sum_{i=1}^{m_1} (a_i - p_i(x^m))^2 = 0 \qquad (1\text{-}21)$$

and that

$$\lim_{m \to \infty} \frac{1}{w^m} \sum_{j=1}^{m_2} \frac{1}{(b_j - q_j(x^m))} = 0 \qquad (1\text{-}22)$$

Thus,

$$\lim_{m \to \infty} A(x^m, w^m) - f(x^*) = 0 \qquad (1-23)$$

In effect, the influence of the constraints on the augmented function is relaxed and finally removed in the limit, and the augmented function converges to the same maximal value as the original objective function.

The function $A(x, w^m)$ of (1-16) is called a mixed <u>interior-exterior-point</u> penalty function. If there were no equality constraints of the form (1-2), then (1-16) would have the form

$$f(x) - \frac{1}{w} \sum_{j=1}^{m_2} \frac{1}{(b_j - q_j(x))} \qquad (1-24)$$

and would be called an <u>interior-point</u> penalty function allowing only feasible points with respect to the inequality constraints. On the other hand, if there were no inequality constraints of the form (1-3), then (1-16) would have the form

$$f(x) - w \sum_{i=1}^{m_1} (a_i - p_i(x))^2 \qquad (1-25)$$

1.6 PRELIMINARY CONCEPTS 35

and would be called an <u>exterior-point</u> penalty function. Although either exterior or interior penalty methods can be used to treat inequality constraints (an exterior method is used in subsequent chapters), interior penalty methods cannot be used to treat equality constraints. Initially, the exterior penalty methods allow points not feasible to the constraints, but force convergence to a feasible point in the limit as $m \to \infty$.

The development and use of sequential unconstrained methods has a long history. The mainstream of developments are reviewed in [F6] and in references cited there. Important results of many of the methods can be found in [A1, B8, F3, F4, F5, L2, L7, M1, P4, W2, Z2].

The usual penalty methods for solving nonlinear programming problems are subject to numerical instabilities, because the derivatives of the penalty functions or the penalty functions themselves increase without bound near the solution as computation proceeds, and are held in check by penalty weighting factors which, to insure convergence, must either increase to infinity or decrease to zero. It is noted [K2] that penalty function methods can become computationally ineffective if certain constraints tend to dominate the entire constraint set. Several authors [L1, L6, M10, P1] confirm this fact by showing that ill-conditioning can occur in the Hessian matrix in the final phases of a penalty function search, although a method to avoid this situation is given in [L6]. When problems do arise in a penalty function search, they invariably occur during the final phase of the search. It is in this phase, however, that relationships between Lagrange multipliers and certain penalty terms become viable, as is shown in [B5] and [F6] for example.

In past years, the idea has arisen that terminal search instabilities might be circumvented by an approach involving a Lagrangian function containing additional penalty-like terms. Most of the work in this direction has been for problems treating equality constraints. The convergence proofs for these new methods often pertain to a class of functions which exhibit appropriate convex properties. Only recently has this type of method been extended to account for inequality constraints and

1.6 PRELIMINARY CONCEPTS 37

to deal effectively with the mixed equality and inequality constraints of the general NLP.

1.7 BRIEF HISTORY OF MULTIPLIER METHODS

Although Courant [C6] formalized the use of penalty functions for nonlinear programming in 1943, it was the work of Fritz John [J1] in 1948 and the fundamental theoretical paper [K6] of Kuhn and Tucker in 1951 that placed nonlinear programming on a substantial foundation. The results of Kuhn and Tucker on necessary conditions and sufficient conditions characterized the solution of the nonlinear convex programming problem and gave an equivalence between this problem and the saddle point problem of the corresponding Lagrangian function. A <u>Lagrangian function</u> L for the NLP of equations (1-1), (1-2), and (1-3) is

$$L \triangleq f + \sum_{i=1}^{m_1} \alpha_i (a_i - p_i)$$
$$+ \sum_{j=1}^{m_2} \beta_j (b_j - q_j) \qquad (1-25)$$

The α_i's and β_j's are real variables and are called <u>Lagrange multipliers</u> (the β_j's are sometimes called generalized Lagrange multipliers). The role that this Lagrangian function plays in optimality conditions is the subject of Chapters 2 and 4. It is known that the existence of a saddle point of the Lagrangian function is heavily dependent upon convexity properties of the underlying problem. In 1956, Arrow and Hurwicz [A6] showed that convexity assumptions could be relaxed by using a modified Lagrangian approach. Differential gradient schemes by Arrow and Solow [A7] in 1958 advanced additional saddle point results in terms of a different modified Lagrangian function. These early modified Lagrangian functions formed the cornerstones of what subsequently developed into augmented Lagrangian algorithms (also called multiplier algorithms). King in 1966 [K3], by relinquishing the convexity property and thus sacrificing the Kuhn-Tucker global properties, gave local necessary conditions for inequality constrained extreme values.

The method of multipliers for (equality constrained) nonlinear programming was independently introduced in 1968 by Hestenes [H3, H4] and Powell

1.7 PRELIMINARY CONCEPTS 39

[P12]. They proposed a dual method of solution in which squares of the constraint functions are added to the Lagrangian. A series of unconstrained minimizations on the penalized Lagrangian is followed by a multiplier vector update according to a simple rule. Powell [P12] showed that if second-order sufficient conditions for optimality are satisfied, the algorithm should converge locally at a linear rate, <u>without requiring the penalty factors to grow without bound</u>. The main advantage of the algorithm lies in the latter property, since it provides a numerical stability that is not found in the usual penalty methods. Miele et al. [M3, M4, M6, M7, M8] have modified and improved the original methods of Hestenes and Powell and have obtained many computational results for the equality constraint case. Haarhoff and Buys [H1] advanced a similar method in 1970.

Fletcher [F8, F9] advanced yet another technique related to that of Hestenes and Powell. Instead of updating the multiplier vector after a sequence of minimizations, the multiplier was adjusted continuously as the minimization on the penalized Lagrangian was performed. Suggestions were given as to how this method could be extended to include inequality constraints. Termination and convergence properties were also given in [F8].

In 1970, Rockafellar [R1] introduced an augmented Lagrangian for inequality constrained convex programming problems for which an unconstrained saddle point corresponded to a solution of the convex programming problem. This approach was further studied by Rockafellar in a series of papers [R2, R4, R5, R6]. Kort and Bertsekas [K4, K5] treat both equality and inequality constraints in their multiplier methods for convex programming.

Arrow, Gould, and Howe [A5] in 1971 studied Rockafellar's augmented Lagrangian, included in a general class of augmented Lagrangians, for nonconvex programming problems and established local saddle point properties for this class of problems. Related approaches to various classes of augmented Lagrangians have been treated by Mangasarian [M2] and others [E2, K7, S2]. The inequality Lagrange multipliers of Rockafellar and Mangasarian are not constrained explicitly to be nonnegative, as they are in this work, but become nonnegative at

1.7 PRELIMINARY CONCEPTS 41

constrained optimal points. Various Lagrangian functions cited above are listed in Appendix C.

In 1971 Pierre [P5] suggested widely applicable multiplier algorithms, accounting for both equality and inequality constraint functions. Pierre [P7, P8] introduced a particularly interesting augmented Lagrangian with local convergence properties; numerical results from an operational algorithm were given. This method was further studied and modified by Lowe [L4].

1.8 INTERLUDE

The literature review in Sections 1.6 and 1.7 is modest in scope. In particular, we have not attempted to review the many interrelated branches of mathematical programming, such as linear programming, integer programming, mixed integer programming, separable programming, dynamic programming, etc. The literature cited does show much of the background effort that has culminated in the construction of highly effective algorithms which generally excel when applied to NLP's of a continuous nature. The functions of Sections 1.3 and

1.4 are representative of the type of functions of interest.

For many NLP's, the convexity conditions of Section 1.5 are not applicable; but whether or not they are applicable, we are interested in optimality conditions (Chapter 2) that characterize constrained local optimal points. These optimality conditions lead in a natural way to the introduction and understanding of Lagrange multipliers and to the role that multipliers play in sensitivity analysis (Chapter 3).

In Section 1.6, methods of sequential unconstrained maximization are considered. In Chapter 4, the scope of these methods is expanded by using augmented objective functions that include Lagrange multipliers and account for the optimality conditions of Chapter 2. The role that local duality plays in this method of multipliers is also investigated. In Chapter 5 and Appendix E, numerical methods are given to implement the various parts of multiplier algorithms. Computational results for a variety of test problems and application problems are given in Chapters 6 and 7.

PROBLEMS PRELIMINARY CONCEPTS

PROBLEMS

1.1 Company Z has to produce 10,000 units of a product during a given time period. It has three plants (numbered 1, 2, and 3) each of which has a limited capacity to produce units of the product during the time period. The number of units (as a function of production cost x_i) that the ith plant can produce during the time period is characterized by Figure 1-3, with a and b replaced by appropriate a_i and b_i constants. A minimum of total production cost is desired. Give the corresponding NLP.

1.2 An architect is to design a simple rectangular storage building having a storage capacity of 1,500 cubic meters. The cost of flooring-plus-land per square meter is estimated to be three times the cost of the walls per m². The cost of the ceiling per m² is estimated to be 1.5 times the cost of walls per m². Let x_1 = width, x_2 = height, and x_3 = length. For esthetic reasons, the front of the building is to satisfy the classical golden section ratio [C8]; that is, $(x_1/x_2) = (x_1 + x_2)/x_1$. The total cost of the building is to be minimized, subject to the above conditions. Give the versions of (1-1) through (1-4) that apply to this NLP.

1.3 The output z of a process depends on two parameters r_1 and r_2. Real values of z are known for 25 different settings of r_1 and r_2, say, for all possible combinations of $r_1 = 1, 2, 3, 4, 5$ and $r_2 = 1, 2, 3, 4, 5$. The output is to be approximated by the function $y = k_0 + k_1 r_1 + k_2 r_2 + k_3 r_1^2 + k_4 r_2^2 + k_5 r_1 r_2$ where k_0 through k_5 are to be selected. It is given that the center value $y(3,3)$ must equal $z(3,3)$ and that

$$\sum_{r_2=1}^{5} \sum_{r_1=1}^{5} [y(r_1,r_2) - z(r_1,r_2)]^2$$

is to be minimized. Formulate the corresponding NLP. (This is a least-squares problem and a quadratic programming problem.)

1.4 A manager has a sum of money to invest and plans to divide it among five ventures. Let x_i represent the percentage of the money to be allocated to the ith venture -- the sum of the x_i's must then equal 100. The expected value of the investment in the ith venture after one year can be characterized by $\bar{r}_i x_i$, where $\bar{r}_i$ is a mean value. (The r_i's are independent normally distributed random variables and are characterized by mean values $\bar{r}_i$'s and variances $\sigma_{r_i}^2$'s.) A maximum of the total expected value is desired; but the manager desires to be 95% confident that the actual value of his investment will be at least 105% at year's end. (Statistical concepts that pertain to this problem are considered in Section 7.3.) Give the corresponding NLP.

1.5 Company Z plans to borrow x dollars this year at interest rate r_1 to expand facilities and production at two plants. Z will pay the money back in equal yearly installments over the next 10 years; the annual payment to the lending agency is therefore γx where $\gamma = r_1(1 + r_1)^{10}/[(1 + r_1)^{10} - 1]$. Of the x dollars, x_1 is for plant 1, and x_2 is for plant 2; $x_1 + x_2 \leq 10^7$. The financial return that will be generated in the kth year from the additional plant capacity is estimated to be $g_1(x_1,k)$ for plant 1 and $g_2(x_2,k)$ for plant 2 (k = 1, 2, ..., 10); $g_1(x_1,10)$ and $g_2(x_2,10)$ include salvage values for the expanded facilities. Company Z can obtain an interest rate of r_2 on money it earns. To be maximized is the total present worth of the proposed expansion -- the present worth of the kth year's net is $[g_1(x_1,k) + g_2(x_2,k) - \gamma(x_1 + x_2)]/(1 + r_2)^k$. Give the NLP.

PROBLEMS   PRELIMINARY CONCEPTS   45

1.6   Consider Figure 1-P6.   The straight lines represent roller-type conveyors; ℓ and h are fixed lengths.   A mass M starts from rest at the top and, with negligible friction, slides to the bottom.   A minimum transision time T from top to bottom is desired.   To be selected to obtain the minimum time T are ℓ_1, θ_1, ℓ_2, and θ_2.   Using physics, derive the relationships that are required to pose this problem as an NLP.   (Partial solution: minimize $T = T_1 + T_2$ where $\ell_1 = (T_1^2)(\sin \theta_1)/(2g)$, and $\ell_2 = [(T_2^2)(\sin \theta_2)/(2g)] + [T_1(\sin \theta_1)/g]T_2$.)

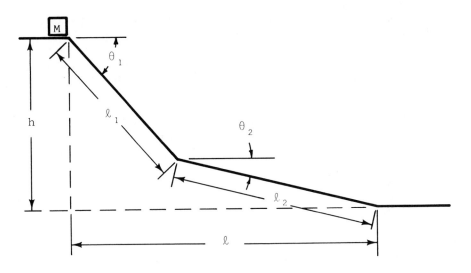

Figure 1-P6   A conveyor problem.

1.7   A given electrical power source is characterized by 10 volts in series with a nonlinear resistor: the source resistance R_s equals $(10 + x_1)$ ohms where x_1 is the current delivered from the source.   We wish to determine the load resistance x_2 which, when attached to the source, results in maximum power transfer from the source to x_2.   From physics, power to the load equals load resistance times the square of load current, where current satisfies Ohm's law.   Give the corresponding NLP.

1.8 What is the NLP formulation for finding the largest circle whose center is at some point (x_{1c}, x_{2c}) to be determined such that the circle is inscribed within the triangle formed by the points (0, 0), (0, 8), and (6, 0)? A similar problem is considered in [G3].

1.9 An outboard motor company has ℓ factories located geographically in metropolitan manufacturing areas and m retail market centers located near popular water recreation areas. Each factory i has a maximum capacity of supplying a_i units per month, and each market center j has a known demand for b_j units per month. It is required to determine the number of units to be transported from each factory to each market center so as to minimize transportation costs while satisfying supply and demand restrictions. It is assumed that the total monthly supply of the factories will satisfy the total monthly demand of the market centers. The transportation cost of each motor from factory i to market center j is c_{ij}. Derive the NLP for this problem in the standard form given in Section 1.2. Is this a linear problem? Is the constraint region concave or convex? Does the problem have but one constrained local optimum value which is the global optimum?

1.10 Let S be the set defined by $x_2 \leq 1.5$, $(x_1 - 1)^2 + (x_2 - 1)^2 \leq 1$, and $(x_1 - 2)^2 + (x_2 - 2)^2 \geq 1$. Show S in the x_1, x_2 plane. If $f(x) = x_1 + x_2$, determine which of the following points are points of local maximum of f over the set S (see Definition 1.2): a) $x_1 = 2$, $x_2 = 1$; b) $x_1 = x_2 = 2 - (1/2)^{\frac{1}{2}}$; c) $x_1 = 2 - (3/4)^{\frac{1}{2}}$, $x_2 = 1.5$; d) $x_1 = x_2 = 1 - (1/2)^{\frac{1}{2}}$. Also, what is the constrained maximum value of f?

PROBLEMS   PRELIMINARY CONCEPTS   47

1.11   Which of the following characterize closed convex sets?

a. $3x_1^2 + x_1 x_2 + x_2^2 \leq 6$

b. $5(1 - e^{-x_1}) + (1/x_2) \geq 2$ and

   $x_1, x_2 \geq 0$

c. $x_2 \geq \sin x_1$ and $x_1, x_2 \geq 0$

d. $x_2 \leq \sin x_1$ and $x_2 \geq (x_1 - 3)^2$

e. $f(x) = c$ where c is a scalar constant and $f(x)$ is any nonlinear function of the vector x. (Can you find a nonlinear $f(x)$ that will result in a convex set being defined by $f(x) = c$?)

1.12   Which of the following functions are concave? Which of them are convex?   Give proofs.

a. $f(x) = x_1(1 - e^{-\gamma x_2})$

   with $x_1, x_2 \geq 0$ and $\gamma > 0$

b. $f(x) = x_1 e^{\gamma x_2}$ with $x_1, x_2 \geq 0$ and

   $\gamma > 0$

c. $f(x) = c + b'x + \frac{1}{2}x'Ax$ with

   A being a positive semidefinite matrix

d. $f(x) = (1/x_1) + (1/x_2)$ with $x_1, x_2 \geq 0$

1.13 Let $f_1(x)$ and $f_2(x)$ be convex functions of the single variable x within the region $c \leq x \leq d$. It was shown by Theorem 1.1 that $f(x) = f_1(x) + f_2(x)$ is also convex in the same region. Is this also always true of the product $p(x)$?

$$p(x) = f_1(x) f_2(x), \quad c \leq x \leq d$$

Give a proof or a counter example.

1.14 A problem that is often encountered is the solution of n simultaneous nonlinear (or linear) algebraic equations:

$$p_i(x) = a_i, \quad i = 1, 2, \ldots, n, \quad x \in E^n$$

How might this problem be transformed into an equivalent constrained NLP? An unconstrained NLP? What would be the advantages or disadvantages of each formulation, given that the solution method required the use of first derivatives?

2
CHARACTERIZATION OF CONSTRAINED MAXIMA: CONDITIONS OF OPTIMALITY

2.1 INTRODUCTION

In a particular situation, one is usually interested in the global solutions to a mathematical programming problem. But in general, only theorems concerning local solutions can be readily proved, unless the problem has certain properties that can be exploited. One such property is that of convexity: as observed in Chapter 1, the convex programming problem involves minimizing a convex function or maximizing a concave function in a convex feasible region, and any local solution is a global solution. For most NLP's, however, maximization algorithms converge to various constrained local maxima depending upon initial search conditions, and global maxima are obtained by comparing the constrained local maxima. Useful theorems can be developed stating necessary conditions and sufficient conditions that a point be a constrained local maximum point to the general NLP. This chapter is devoted to the presentation and illustration of such conditions with regard to the Lagrangian function that is associated with the NLP.

2.2 ASSUMPTIONS AND DEFINITIONS

Consider again the general constrained problem introduced in Section 1.2. Find x^* and $f(x^*)$ where

$$f(x^*) = \max_x f(x) \qquad (2-1)$$

subject to the constraints

$$p_i(x) = a_i, \quad i = 1, 2, \ldots, m_1 < n \qquad (2-2a)$$

and

$$q_j(x) \leq b_j, \quad j = 1, 2, \ldots, m_2 \qquad (2-3a)$$

Constraints (2-2a) and (2-3a) are expressed more compactly as

$$p(x) = a \qquad (2-2b)$$

2.2 CONDITIONS OF OPTIMALITY

and

$$q(x) \leq b \qquad (2-3b)$$

where p, q, a, and b are column vectors of appropriate dimensions; x too is expressed as a column vector.

Let x^* be a point which satisfies (2-2) and (2-3), and let Δx denote an incremental change in x from x^*, i.e., $\Delta x = x - x^*$, where the magnitude $\|\Delta x\|$ is the Euclidean norm† $(\Delta x' \Delta x)^{\frac{1}{2}}$. The following definition is closely related to Definition 1.2: the closed set S of Definition 1.2 is represented here by $\{x \mid p(x) = a \text{ and } q(x) \leq b\}$.

DEFINITION 2.1 [Constrained Local Maximum]
A function f is said to have a <u>constrained local maximum</u> at x^* if there exists a real number $\varepsilon > 0$ such that

$$f(x^*) \geq f(x^* + \Delta x) \qquad (2-4)$$

---

† The symbol ´ is used to denote the transpose of a vector or matrix. Let C be an m×n matrix. Then C´ is the transpose of C and has dimension n×m where the (i,j)th entry of C´ is c_{ji}.

for all Δx in the set

$$\{\Delta x \mid 0 < \|\Delta x\| < \varepsilon;\ p_i(x^* + \Delta x) = a_i,$$
$$i = 1, 2, \ldots, m_1;\ \text{and}\ q_j(x^* + \Delta x) \leq b_j$$
$$j = 1, 2, \ldots, m_2\}$$

If (2-4) in the above definition can be replaced by

$$f(x^*) > f(x^* + \Delta x) \qquad (2-5)$$

then f is said to have a <u>strict constrained local maximum</u> at x^*.

In the theorems of this chapter, we assume that the objective function and the constraint functions of the NLP are continuous and exhibit a suitable degree of differentiability. Also, the n-dimensional gradient vectors $\nabla p_i(x)$ and $\nabla q_j(x)$ are assumed to be nonzero at those points x which satisfy $p_i(x) = a_i$ and $q_j(x) = b_j$, respectively. To be precise, the following definitions are in order.

DEFINITION 2.2 [Continuity]
A real valued function f, defined on a set X in

2.2 CONDITIONS OF OPTIMALITY 53

E^n, is said to be <u>continuous at a point</u> $x_0 \in X$ if, given any number $\varepsilon > 0$, there is a nonempty neighborhood N_{ε/x_0}, such that $|f(x) - f(x_0)| < \varepsilon$ for every point $x \in N_{\varepsilon/x_0} \subset X$. The function f is said to be <u>continuous</u> on X if it is continuous at each point of X.

DEFINITION 2.3   [Class of Differentiability]
A function f is said to be of class C^0 in an open region X in E^n if f is continuous in X. It is said to be of class C^1 in X if it is of class C^0 in X and its first-order partial derivatives with respect to x, $\partial f/\partial x_i$, i = 1, 2, ..., n, are defined and continuous in X. It is said to be of class C^2 in X if it is of class C^1 and its second-order partial derivatives, $\partial^2 f/\partial x_i x_j$, i,j = 1, 2, ..., n, are defined and continuous in X.

Differentiability of higher order C^k, k > 2, is defined in similar fashion, but is not required in this work.

DEFINITION 2.4   [Differentiability over an Extended Feasible Region]
Let $S_\nu \triangleq \{x | q(x) < b + \nu,$ and $a - \nu < p(x) < a + \nu\}$

where $\nu > 0$. The NLP functions are said to be of class C^k over an extended feasible region if they are of class C^k in S_ν for some vector $\nu > 0$.

The first-order conditions (second-order conditions) in this chapter are generally applicable to NLP's which have functions that are of class C^1 (C^2) over an extended feasible region.

2.3 TAYLOR SERIES IN n VARIABLES

The Taylor series is a useful tool in the development of optimality conditions. Consider a real-valued function $f(x)$ of n variables which are expressed in column vector form as $x = [x_1 \ x_2 \ \ldots \ x_n]'$. Let x^* be a particular x vector of interest and define $\xi \Delta x \triangleq x - x^*$ where ξ is a real scalar.

Thus,

$$x = x^* + \xi \Delta x \qquad (2\text{-}6)$$

and

$$\frac{dx}{d\xi} = \Delta x = [\Delta x_1 \ \Delta x_2 \ \ldots \ \Delta x_n]' \qquad (2\text{-}7)$$

2.3 CONDITIONS OF OPTIMALITY

Assume that f is of class C^2 and that, for the moment, both x^* and Δx are fixed; then

$$f(x) = f(x^* + \xi \Delta x) \triangleq y(\xi) \qquad (2-8)$$

where $y(\xi)$ is a parametric representation of $f(x)$, and $y(\xi)$ can be expanded in a MacLaurin series as follows:

$$y(\xi) = y(0) + \frac{dy(0)}{d\xi} \xi + \tfrac{1}{2} \frac{d^2 y(0)}{d\xi^2} \xi^2 + \cdots \qquad (2-9)$$

The terms in (2-9) depend upon f; that is,

$$y(0) = f(x^*) \qquad (2-10)$$

$$\frac{dy(0)}{d\xi} = \sum_{i=1}^{n} \frac{\partial f(x^*)}{\partial x_i} \frac{dx_i}{d\xi} = \nabla f(x^*)' \Delta x \qquad (2-11)$$

and

$$\frac{d^2 y(0)}{d\xi^2} = \sum_{i=1}^{n} \sum_{j=1}^{n} \frac{\partial^2 f(x^*)}{\partial x_i \partial x_j} \frac{\partial x_i}{\partial \xi} \frac{\partial x_j}{\partial \xi}$$

$$= \Delta x' \nabla^2 f(x^*) \Delta x \qquad (2-12)$$

where $\nabla f(x^*)$ is the gradient of $f(x)$ evaluated at x^*, and $\nabla^2 f(x^*)$ is the $n \times n$ <u>Hessian</u> matrix of f evaluated at x^*. The $(i,j)^{th}$ entry of the Hessian is $\partial^2 f(x^*) / \partial x_i \partial x_j$.

Of particular interest is the form of (2-9) with $\xi = 1$. Letting $\xi = 1$ and substituting (2-10), (2-11), and (2-12) into (2-9), we have the Taylor series

$$f(x) = f(x^*) + \Delta x' \nabla f(x^*) + \tfrac{1}{2} \Delta x' \nabla^2 f(x^*) \Delta x$$
$$+ \, 0(\|\Delta x\|^3) \qquad (2\text{-}13)$$

where the remainder $0(\|\Delta x\|^3)$ has the property that

$$\lim_{\|\Delta x\| \to 0} \left[\frac{0(\|\Delta x\|^3)}{\|\Delta x\|^2} \right] \to 0$$

In (2-13), the components of Δx are the variables and are related to x by $x = x^* + \Delta x$; the Δx_i's play the same role in (2-13) that ξ plays in the MacLaurin series (2-9).

2.4 CONDITIONS FOR UNCONSTRAINED MAXIMA AND MINIMA

Consider the following rearrangement of (2-13):

$$f(x) - f(x^*) = \Delta x' \nabla f(x^*) + \cdots \qquad (2\text{-}14)$$

If $\nabla f(x^*) \neq 0$, there exists an arbitrarily small $\|\Delta x\|$ with the property that both

$$f(x^* + \Delta x) - f(x^*) \approx \Delta x' \nabla f(x^*) > 0 \qquad (2\text{-}15)$$

2.4 CONDITIONS OF OPTIMALITY

and

$$f(x^* - \Delta x) - f(x^*) \cong -\Delta x' \nabla f(x^*) < 0 \qquad (2\text{-}16)$$

in which case, x^* has neighboring values that yield both larger and smaller values of f. Therefore, $\nabla f(x^*)$ must be zero at an unconstrained optimum. Thus, the following theorem holds.

THEOREM 2.1 [First-order Necessary Condition for an Unconstrained Local Optimum]
A necessary condition that a function f of class C^1 have an unconstrained local maximum (minimum) at a point x^* is that

$$\nabla f(x^*) = 0 \qquad (2\text{-}17)$$

Condition (2-17) gives a set of n equations in n unknowns, the x_i^* values. Because the equations are nonlinear, in general, they may exhibit any number of solutions, including the possibility of no solution. Once a solution is obtained, however, it should be examined in the light of sufficient conditions.

THEOREM 2.2 [Second-order Sufficient Condition for an Unconstrained Local Optimum]

A sufficient condition that a function f of class C^2 have an unconstrained local maximum (minimum) at a point x^* is that (2-17) holds and that $\Delta x' \nabla^2 f(x^*) \Delta x < 0$ ($\Delta x' \nabla^2 f(x^*) \Delta x > 0$) for all $\Delta x \neq 0$.

Proof. With $\nabla f(x^*) = 0$, the Taylor series (2-13) reduces to

$$f(x) - f(x^*) = \tfrac{1}{2} \Delta x' \nabla^2 f(x^*) \Delta x + O(\|\Delta x\|^3) \quad (2\text{-}18)$$

Consider the open neighborhood $\|x - x^*\| < \varepsilon$ for some real number $\varepsilon > 0$. We assume that $O(\|\Delta x\|^3)$ is negligible for $\|\Delta x\| < \varepsilon$. If $\Delta x' \nabla f(x^*) \Delta x$ is less than (greater than) zero for any $\Delta x \neq 0$, it must follow that $f(x^*)$ is an unconstrained local maximum (minimum) of $f(x)$.

●

Theorem 2.2 is closely linked to a second-order necessary condition (Problem 2.2). Namely, we must replace "sufficient" in the theorem by "necessary" if the strict inequalities are replaced by $\Delta x' \nabla^2 f(x^*) \Delta x \leq 0$ ($\Delta x' \nabla^2 f(x^*) \Delta x \geq 0$).

2.4 CONDITIONS OF OPTIMALITY

In terms of matrix theory, $\Delta x' \nabla^2 f(x^*) \Delta x$ is less than (greater than) zero for all $\Delta x \neq 0$ if $\nabla^2 f(x^*)$ is a <u>negative</u> (positive) <u>definite matrix</u>. Systematic tests for positive definiteness and negative definiteness are readily available (Appendix D).

2.5 CASE OF ONE EQUALITY CONSTRAINT

Consider the equality constraint

$$p_1(x) = a_1 \tag{2-19}$$

where $p_1(x)$ is assumed to be a function of class C^2 and a_1 is a given constant. A maximum of $f(x)$, subject to satisfying (2-19), is desired. Let x^* be a constrained local maximum point and consider the following Taylor series:

$$f(x) - f(x^*) = \Delta x' \nabla f(x^*) + \tfrac{1}{2} \Delta x' \nabla^2 f(x^*) \Delta x + \cdots \tag{2-20}$$

and

$$p_1(x) - p_1(x^*) = \Delta x' \nabla p_1(x^*) + \tfrac{1}{2} \Delta x' \nabla^2 p_1(x^*) \Delta x + \cdots \tag{2-21}$$

where both $p_1(x)$ and $p_1(x^*)$ equal a_1; that is, allowed values of x are required to satisfy $p_1(x) = a_1$, and therefore, (2-21) imposes the following constraint on allowed Δx's.

$$0 = \Delta x' \nabla p_1(x^*) + \cdots \qquad (2\text{-}22)$$

To first-order terms, equation (2-22) is

$$(x - x^*)' \nabla p_1(x^*) = 0 \qquad (2\text{-}23)$$

which defines a tangent hyperplane[†] to the surface characterized by $p_1(x) = a_1$. A vector that is normal to the plane is $\nabla p_1(x^*)$, and the plane passes through the point x^*.

We know from (2-22) that $\Delta x' \nabla p_1(x^*) = 0$ is required for sufficiently small $\|\Delta x\|$. Only if $\nabla f(x^*)$ is some scalar multiple of $\nabla p_1(x^*)$, say $\nabla f(x^*) = \alpha_1 \nabla p_1(x^*)$, will equation (2-20) reduce to

$$f(x) - f(x^*) = \tfrac{1}{2} \Delta x' \nabla^2 f(x^*) \Delta x + \cdots \qquad (2\text{-}24)$$

in which case $f(x^*)$ could conceivably be a constrained maximum. Thus, a necessary condition that

---

[†] A hyperplane is a plane in n dimensions.

2.5 CONDITIONS OF OPTIMALITY

$f(x^*)$ be a constrained local maximum is that

$$\nabla f(x^*) - \alpha_1^* \nabla p_1(x^*) = 0 \qquad (2\text{-}25)$$

Let

$$L(x) \triangleq f(x) + \alpha_1 [a_1 - p_1(x)] \qquad (2\text{-}26)$$

$L(x)$ is a Lagrangian function, and α_1 is a Lagrange multiplier. Technically, this Lagrangian should be denoted by $L(x, \alpha_1)$ to express its dependence on the multiplier. For notational convenience, however, $L(x)$ or simply L is often used.

In terms of $L(x)$, (2-25) is compactly expressed as

$$\nabla L(x^*) = 0 \qquad (2\text{-}27)$$

Condition (2-27) and constraint (2-19) constitute $n + 1$ equations in $n + 1$ unknowns, the n x_i^*'s and α_1^*. Any number of solutions to these equations is possible, depending on f, p_1, and a_1. A given solution should be tested by applying a sufficient condition for local maxima.

To obtain the second-order sufficient condition that applies to this problem, we multiply

(2-21) by the multiplier α_1, and then subtract the result from (2-20). Thus, assuming (2-19) and (2-25) hold,

$$f(x) - f(x^*) = \tfrac{1}{2}\Delta x' \nabla^2 L(x^*) \Delta x + \cdots \qquad (2-28)$$

From this it is clear that a sufficient condition for $f(x^*)$ to be a constrained local maximum is that $\nabla L(x^*) = 0$ and $\Delta x' \nabla^2 L(x^*) \Delta x < 0$ for any $\Delta x \neq 0$ which satisfies $\Delta x' \nabla p_1(x^*) = 0$.

EXAMPLE 2.1 Figure 2-1 gives a geometrical interpretation of the multiplier α_1 in the case that $n = 2$ with one equality constraint. Along the set of points satisfying $p_1(x) = a_1$, most points are like x^0 in the sense that $\nabla f(x^0)$ is not a scalar multiple of $\nabla p_1(x^0)$. Points x^1, x^2, x^3, and x^4 are the exceptions in Figure 2-1; at these four points, $\nabla f(x)$ is a scalar multiple of ∇p_1. The $f(x^4) = 3$ value is clearly a constrained local maximum; $f(x^3) = 1.5$ is a constrained local minimum; $f(x^2) = 5$ is both a constrained local maximum and the absolute constrained maximum for the problem; and $f(x^1) = 2$ corresponds to a saddle point, having

2.5 CONDITIONS OF OPTIMALITY

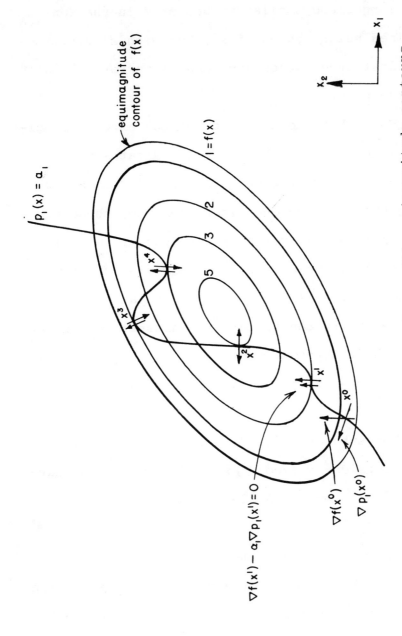

Figure 2-1   Contour of $p_1(x) = a_1$ along with equimagnitude contours of $f(x)$.

both larger and smaller values of f in the constrained neighborhood of x^1, but with x^1 satisfying first-order necessary conditions for a local optimum point.

Note that the direction of the arrows associated with $\nabla p_1(x)$ in Figure 2-1 indicates that $p_1(x) > a_1$ above the curve and $p_1(x) < a_1$ below the curve. If the opposite were true, these arrows would simply point in the opposite direction, meaning that the multipliers that yield $\nabla L = 0$ would change sign. The multipliers associated with a given equality constraint are unrestricted in sign at a constrained optimum.

In order to illustrate more clearly the conditions that must be satisfied when maximizing a function subject to one equality constraint, we shall consider a second example which can be carried to completion in closed form.

EXAMPLE 2.2 [Area-Volume Problem]
Maximize the volume V of a cylindrical tank which has an open top, subject to the constraint that the total surface area A is a constant. The

2.5 CONDITIONS OF OPTIMALITY

surface area consists of the wall and the bottom of the tank. Let x_1 denote the radius of the cylinder and x_2 its height; also let $f(x_1, x_2)$ equal the normalized volume V/π; and let $p_1(x_1, x_2) = A/2\pi = a_1$ (a constant) be the normalized area. We have then

$$f(x_1, x_2) = x_1^2 x_2 \qquad (2\text{-}29)$$

and

$$p_1(x_1, x_2) = x_1 x_2 + (x_1^2/2) = a_1 \qquad (2\text{-}30)$$

The Lagrangian L is

$$L = x_1^2 x_2 + \alpha_1 (a_1 - x_1 x_2 - (x_1^2/2)) \qquad (2\text{-}31)$$

for which the gradient is

$$\nabla L = \begin{bmatrix} 2x_1 x_2 - \alpha_1 (x_2 + x_1) \\ \\ x_1^2 - \alpha_1 x_1 \end{bmatrix} \qquad (2\text{-}32)$$

At $x = x^*$ and $\alpha_1 = \alpha_1^*$, $\nabla L^* = 0$ can be shown to give

$$\begin{bmatrix} x_1^* \\ x_2^* \end{bmatrix} = \begin{bmatrix} \alpha_1^* \\ \alpha_1^* \end{bmatrix} \qquad (2\text{-}33)$$

And from (2-30) and (2-33),

$$\alpha_1^{*2} + (\alpha_1^{*2}/2) = a_1 \qquad (2-34)$$

or

$$\alpha_1^* = (2a_1/3)^{\frac{1}{2}} \qquad (2-35)$$

Consider $\nabla^2 L^*$,

$$\nabla^2 L^* = \begin{bmatrix} 2x_2^* - \alpha_1^* & 2x_1^* - \alpha_1^* \\ \\ 2x_1^* - \alpha_1^* & 0 \end{bmatrix} = \alpha_1^* \begin{bmatrix} 1 & 1 \\ 1 & 0 \end{bmatrix} \qquad (2-36)$$

Note that $-\nabla^2 L^*$ is not a positive-definite matrix; therefore, this example illustrates the fact that the Lagrangian need not be maximized at the constrained maximum point.

However, consider any nonzero Δx that satisfies $\Delta x' \nabla p_1^* = 0$,

$$[\Delta x_1 \; \Delta x_2] \begin{bmatrix} 2\alpha_1^* \\ \\ \alpha_1^* \end{bmatrix} = (2\Delta x_1 + \Delta x_2)\alpha_1^* = 0 \qquad (2-37)$$

which can be true only if $\Delta x_2 = -2\Delta x_1$. For such a Δx, we have

2.4 CONDITIONS OF OPTIMALITY

$$\Delta x' \nabla^2 L^* \Delta x = (\Delta x_1)^2 \alpha_1^* [1 \ -2] \begin{bmatrix} 1 & 1 \\ 1 & 0 \end{bmatrix} \begin{bmatrix} 1 \\ -2 \end{bmatrix}$$

$$= (\Delta x_1)^2 \alpha_1^* [1 \ -2] \begin{bmatrix} -1 \\ 1 \end{bmatrix}$$

$$= -3\alpha_1^* (\Delta x_1)^2 < 0 \qquad (2-38)$$

And therefore, the second-order sufficient condition is satisfied at $x = x^*$.

2.6 CASE OF SEVERAL INEQUALITY CONSTRAINTS

Consider the inequality constraints

$$q_1(x) \leq b_1 \qquad (2\text{-}39a)$$

$$q_2(x) \leq b_2 \qquad (2\text{-}39b)$$

and

$$q_3(x) \leq b_3 \qquad (2\text{-}39c)$$

where q_j's are functions of class C^2 over an extended feasible region. The function $f(x)$ is to be maximized with respect to x's which satisfy (2-39).

Let x^* denote a point which is suspected to be a constrained local maximum point; and suppose, for example, that $q_1(x^*) = b_1$, $q_2(x^*) = b_2$, but $q_3(x^*) < b_3$. In this case, q_1 and q_2 are said to be <u>active constraint</u> functions at x^*, and q_3 is an <u>inactive constraint</u> function at x^*. In the following development, the Δx's considered are assumed to satisfy $q_3(x^* + \Delta x) < b_3$.

As in the single equality constraint case, the Taylor series of each of the active constraint functions are required:

$$q_1(x) - q_1(x^*) = \Delta x' \nabla q_1(x^*)$$
$$+ \tfrac{1}{2} \Delta x' \nabla^2 q_1(x^*) \Delta x + \cdots \quad (2\text{-}40a)$$

and

$$q_2(x) - q_2(x^*) = \Delta x' \nabla q_2(x^*)$$
$$+ \tfrac{1}{2} \Delta x' \nabla^2 q_2(x^*) \Delta x + \cdots \quad (2\text{-}40b)$$

Only Δx's for which both $q_1(x) - q_1(x^*) = q_1(x) - b_1 \leq 0$ and $q_2(x) - q_2(x^*) = q_2(x) - b_2 \leq 0$ are allowed. Thus, from (2-40), within first-order terms, we require that

$$\Delta x' \nabla q_1(x^*) \leq 0 \quad (2\text{-}41a)$$

2.6 CONDITIONS OF OPTIMALITY

and

$$\Delta x' \nabla q_2(x^*) \leq 0 \qquad (2\text{-}41b)$$

Also within first-order terms,

$$f(x) - f(x^*) \cong \Delta x' \nabla f(x^*) \qquad (2\text{-}42)$$

From (2-42), if $\Delta x' \nabla f(x^*) < 0$ for all nonzero Δx's that satisfy (2-41), then $f(x) - f(x^*) < 0$ for allowed nonzero Δx's, and $f(x^*)$ is clearly a constrained maximum. However, if we are given only that $\Delta x' \nabla f(x^*) \leq 0$ for all Δx's that satisfy (2-41), then $f(x^*)$ may be a constrained maximum, but higher-order terms in the series expansions should be examined to fully ascertain the nature of x^*. With the exception of abnormal cases (Section 2.7), a necessary condition for a constrained local maximum is that

$$\Delta x' \nabla f(x^*) \leq 0 \qquad (2\text{-}43)$$

for all Δx's that satisfy (2-41).

Suppose that $\nabla f(x^*)$ can be constructed from $\nabla q_1(x^*)$, $\nabla q_2(x^*)$, and some real values β_1^* and $\beta_2^* \geq 0$ in the following way:

$$\nabla f(x^*) = \beta_1^* \nabla q_1(x^*) + \beta_2^* \nabla q_2(x^*) \qquad (2\text{-}44)$$

Then

$$\Delta x' \nabla f(x^*) = \beta_1^* \Delta x' \nabla q_1(x^*) + \beta_2^* \Delta x' \nabla q_2(x^*) \quad (2\text{-}45)$$

From (2-45), any Δx which satisfies (2-41) also satisfies (2-43), and therefore, $f(x^*)$ may be a constrained local maximum.

Let the Lagrangian L for this problem be defined as

$$L(x) = f(x) + \sum_{j=1}^{3} \beta_j [b_j - q_j(x)] \qquad (2\text{-}46)$$

where $\beta_3 = 0$. For normal cases, first-order necessary conditions for $f(x^*)$ to be a constrained local maximum are

$$\nabla L(x^*) = 0 \qquad (2\text{-}47a)$$

$$\beta_j^* [q_j(x^*) - b_j] = 0, \quad j = 1, 2, 3 \qquad (2\text{-}47b)$$

and

$$\beta_j^* \geq 0, \quad j = 1, 2, 3 \qquad (2\text{-}47c)$$

The above conditions, and second-order sufficient

2.6 CONDITIONS OF OPTIMALITY

conditions for the problem of this section, are special cases of the optimality conditions that are developed in a general framework in Sections 2.8 through 2.11.

EXAMPLE 2.3 Figure 2-2 gives a geometrical interpretation of the multipliers β_1 and β_2. The necessary conditions (2-47) are equivalent to the statement that $\nabla f(x^*)$ must be in an n-dimensional region defined by $\beta_1 \nabla q_1(x^*) + \beta_2 \nabla q_2(x^*)$ for all β_1, $\beta_2 \geq 0$, where q_1 and q_2 are the active inequality constraints.

EXAMPLE 2.4 [Parcel Problem]
This example is similar to one considered by Rosenbrock [R9]. A rectangular parcel is to be sent by mail. The length must not exceed 1 meter, and the length and girth combined must not exceed 1.8 meters. What dimensions of the parcel allow the greatest volume? Let x_1 denote the length of the parcel, x_2 its width, and x_3 its depth; let $f(x)$ equal the volume.

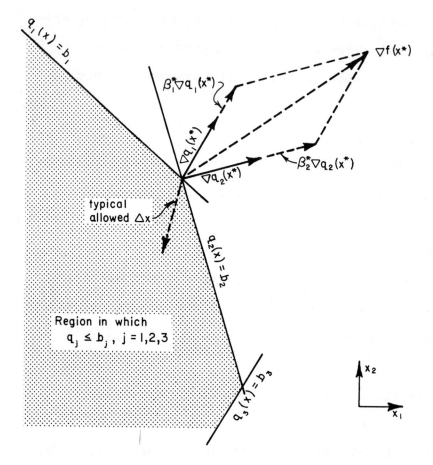

Figure 2-2 Case of several inequality constraints.

We have then

$$f(x) = x_1 x_2 x_3 \tag{2-48}$$

$$q_1(x) = x_1 \leq 1 \tag{2-49}$$

and

2.6 CONDITIONS OF OPTIMALITY

$$q_2(x) = x_1 + 2x_2 + 2x_3 \leq 1.8 \qquad (2\text{-}50)$$

The Lagrangian L is

$$L = x_1 x_2 x_3 + \beta_1 (1 - x_1)$$
$$+ \beta_2 (1.8 - x_1 - 2x_2 - 2x_3) \qquad (2\text{-}51)$$

for which the gradient is

$$\nabla L = \begin{bmatrix} x_2 x_3 - \beta_1 - \beta_2 \\ x_1 x_3 - 2\beta_2 \\ x_1 x_2 - 2\beta_2 \end{bmatrix} \qquad (2\text{-}52)$$

At $x = x^*$ and $\beta = \beta^* \geq 0$, $\nabla L^* = 0$ can be shown to give $(x_1^*, x_2^*, x_3^*) = (0.6, 0.3, 0.3)$ and $(\beta_1^*, \beta_2^*) = (0, 0.09)$, where q_1^* is inactive and q_2^* is an active inequality constraint function.

2.7 ABNORMAL CASE

In Sections 2.5 and 2.6, we developed conditions for $f(x^*)$ to be a local maximum when restricted by one equality constraint and then by several inequality constraints. We found that in both cases a necessary condition for a constrained local maximum is that

$$\Delta x' \nabla f(x^*) \leq 0 \qquad (2-53)$$

for all Δx's that are sufficiently small in magnitude and which result in $x^* + \Delta x$ vectors that satisfy the given constraints. Also, a Lagrangian L was formed using f, the constraints, and appropriate multipliers, such that at x^*,

$$\nabla L^* = 0 \qquad (2-54)$$

In the <u>normal case</u>, (2-54) is valid, and $\nabla f(x^*)$ is a linear combination of the gradients of the active constraints at x^*. <u>Abnormal cases</u> exist, however, for which $\nabla f(x^*)$ cannot be formed as a linear combination of the gradients of the active constraints. In all cases of abnormality, the ∇p_i^*'s and active ∇q_j^*'s exhibit linear dependence, as is evident in the following example.

EXAMPLE 2.5   [An Abnormal Case]
Kuhn and Tucker [K6] give the following single constraint set to illustrate the abnormal case.

$$\begin{aligned} q_1(x) &= x_2 - (1-x_1)^3 \leq 0 \\ q_2(x) &= -x_1 \leq 0 \end{aligned} \qquad (2-55)$$

2.7 CONDITIONS OF OPTIMALITY

and

$$q_3(x) = -x_2 \leq 0$$

Let $f = x_1$ be the function to be maximized. Figure 2-3 depicts the allowed region, and $x^* = (1, 0)$ is clearly evident. At x^*, the active constraints are q_1 and q_3. Note in Figure 2-3 that both Δx^a and Δx^b satisfy $\Delta x' \nabla q_j \leq 0$ for $j = 1, 3$ (here $\Delta x' \nabla q_j^* = 0$ for $j = 1, 3$) but that $\Delta x^{b\prime} \nabla f^*$ is greater than 0 and does not satisfy (2-43). The vectors ∇q_1^* and ∇q_3^* are linearly dependent, and $x^* + \Delta x^b$ is not contained in the allowed region.

Fortunately, abnormal cases rarely occur in practice, and may occur only if the gradients of the active constraints are linearly dependent at x^*. We therefore focus our attention on the normal case.

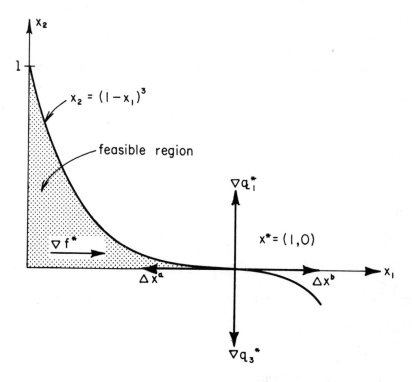

Figure 2-3 An abnormal case.

EXAMPLE 2.6 [A Normal Case]

In the previous example, replace the first constraint by

$$q_1(x) = x_2 - (1 - x_1)^3 \leq 0.125 \qquad (2\text{-}56)$$

with all other conditions of the example unchanged. Note that $q_1(x)$ is the same function and only the right-hand side b_1 has been changed by a small

2.7 CONDITIONS OF OPTIMALITY

amount. Figure 2-4 shows the allowed region in which $x^* = (1.5, 0)$. At x^*, $\nabla q_1^* = (0.75, 1.0)$ and $\nabla q_3^* = (0, -1.0)$. Note from the figure that vectors Δx^a, Δx^b, and any nonnegative linear combination of Δx^a and Δx^b constitute all of the vectors that satisfy $\Delta x' \nabla q_j^* \leq 0$ for $j = 1, 3$; and for all such Δx's, $\Delta x' \nabla f^* \leq 0$ (here in fact $\Delta x' \nabla f^* < 0$). The vectors ∇q_1^* and ∇q_3^* are linearly independent in Figure 2-4.

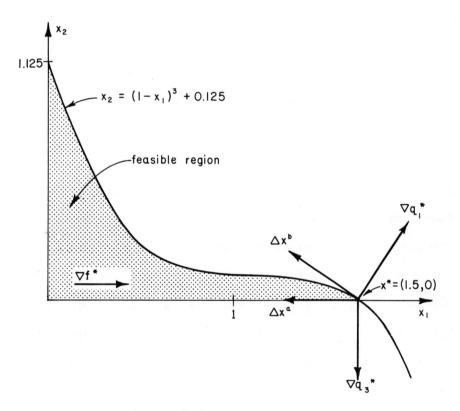

Figure 2-4   A normal case.

We now turn to the general case and to the first- and second-order conditions that apply to a local maximum when constrained by both equalities and inequalities as in the general NLP.

2.8 PRELIMINARY CONDITIONS FOR THE GENERAL CASE

The necessary conditions that apply to the NLP are associated with the Lagrangian L:

$$L(x,\xi,\alpha,\beta) = \xi f(x) + \sum_{i=1}^{m_1} \alpha_i(a_i - p_i)$$

$$+ \sum_{j=1}^{m_2} \beta_j(b_j - q_j) \qquad (2\text{-}57)$$

where α and β are Lagrange multiplier vectors of dimension m_1 and m_2, respectively. The objective function $f(x)$ has an associated multiplier $\xi \geq 0$. The conditions that result for $\xi \geq 0$ are those of Fritz John [J1] which are discussed and implemented in [P8]. For the special case where $\xi = 1$, the resulting conditions are those of Kuhn and Tucker and are presented in the following development.

2.8 CONDITIONS OF OPTIMALITY

With $\xi = 1$, and using matrix notation, equation (2-57) is recast in the form

$$L \triangleq L(x,\alpha,\beta) = f + \alpha'(a - p) + \beta'(b - q) \qquad (2\text{-}58)$$

From (2-58), the gradient of L with respect to x follows as

$$\nabla L = \nabla f - \nabla p' \alpha - \nabla q' \beta \qquad (2\text{-}59)$$

Notation to be used in establishing the necessary conditions that must hold for x^* to be a local maximum is as follows: let f^* denote $f(x^*)$, ∇f^* denote $\nabla f(x^*)$, etc., and let $\Delta x = x - x^*$ denote a change in x that is sufficiently small in magnitude.

At a constrained maximum point x^*, for each active inequality constraint, a series expansion is

$$q_j(x^* + \Delta x) = q_j^* + \Delta x' \nabla q_j^* + \cdots, \quad j \in S_a \qquad (2\text{-}60)$$

where

$$S_a \triangleq \{j \mid q_j^* = b_j\} \qquad (2\text{-}61a)$$

and

$$q_j(x^* + \Delta x) \leq b_j \quad \text{for all } j \qquad (2\text{-}61b)$$

Inequality (2-61b) gives contraints on the allowed Δx's. From (2-60) and (2-61), Δx's must satisfy

$$\Delta x' \nabla q_j^* \leq 0, \quad j \in S_a \tag{2-62}$$

Also, in the neighborhood of x^*,

$$p_i^* = a_i, \quad i = 1, 2, \ldots, m_1 \tag{2-63a}$$

and

$$p_i(x^* + \Delta x) = a_i, \quad i = 1, 2, \ldots, m_1 \tag{2-63b}$$

where (2-63b) further constrains allowed Δx's. A series expansion of $p_i(x^* + \Delta x)$ is

$$p_i(x^* + \Delta x) = p_i^* + \Delta x' \nabla p_i^* + \cdots \tag{2-64}$$

and based on (2-63), this series gives, within first-order terms,

$$\Delta x' \nabla p_i^* = 0, \quad i = 1, 2, \ldots, m_1 \tag{2-65}$$

Equivalently, both

$$\Delta x' \nabla p_i^* \leq 0, \quad i = 1, 2, \ldots, m_1 \tag{2-66a}$$

and

2.8 CONDITIONS OF OPTIMALITY

$$-\Delta x' \nabla p_i^* \leq 0, \quad i = 1, 2, \ldots, m_1 \qquad (2\text{-}66b)$$

are conditions which allowed Δx's must satisfy.

Now, we are interested in the difference

$$f(x^* + \Delta x) - f^* = \Delta x' \nabla f^* + \cdots \qquad (2\text{-}67)$$

By definition, f^* is a constrained maximum if $f(x^* + \Delta x) - f^* \leq 0$ for all Δx's which are sufficiently small in magnitude and which satisfy (2-61b) and (2-63b). If $\Delta x' \nabla f^*$ in (2-67) is strictly less than zero for all Δx's which satisfy (2-62) and (2-65), then the fact that f^* is a constrained local maximum follows directly from the definition. However, if

$$\Delta x' \nabla f^* \leq 0 \qquad (2\text{-}68)$$

for all Δx's which satisfy (2-62) and (2-65), then f^* can conceivably be a constrained local maximum, but higher-order terms in the series expansions have to be investigated to fully ascertain the nature of f^*.

On the basis of the above paragraph, we might be tempted to conclude that <u>all</u> constrained local maxima are characterized by (2-68) being satisfied for all Δx's which satisfy (2-62) and (2-65). However, the abnormal case of Example 2.5 shows that such a conclusion is not justified.

We may now summarize the preceding development. Define the following sets in E^n.

$$S_1^* \triangleq \{\Delta x \mid \Delta x'\nabla q_j^* \leq 0, j \in S_a; \Delta x'\nabla p_i^* = 0,$$
$$i = 1, 2, \ldots, m_1; \Delta x'\nabla f^* \leq 0\} \quad (2\text{-}69)$$

$$S_2^* \triangleq \{\Delta x \mid \Delta x'\nabla q_j^* \leq 0, j \in S_a; \Delta x'\nabla p_i^* = 0,$$
$$i = 1, 2, \ldots, m_1; \Delta x'\nabla f^* > 0\} \quad (2\text{-}70)$$

and

$$S_3^* \triangleq \{\Delta x \mid \Delta x'\nabla q_j^* > 0, \text{ for at least one } j \in S_a;$$
$$\text{or } \Delta x'\nabla p_i^* \neq 0 \text{ for at least one } i\} \quad (2\text{-}71)$$

where x^* is assumed to satisfy (2-2) and (2-3) and where S_a is given in (2-61a).

Clearly, the intersection $S_1^* \cap S_2^* \cap S_3^*$ equals the null set ϕ, and any Δx in E^n belongs to one of the three sets S_1^*, S_2^*, or S_3^*. The set S_1^* of (2-69) is of limited interest because any Δx in this set does not show obvious increases in $f(x)$;

2.8 CONDITIONS OF OPTIMALITY

i.e., $f(x) - f(x^*) \simeq \Delta x' \nabla f^* \leq 0$. Also, the set S_3^* of (2-71) is of limited interest because any Δx in this set does not give an x point that satisfies all of the constraints. The important set is therefore S_2^* of (2-70). A necessary condition that x^* be a constrained local maximum point is that all Δx's contained in S_2^* result in x's which violate one or more of the constraints in (2-2) and (2-3). In the normal case, S_2^* equals the null set ϕ at the constrained maximum point x^*.

2.9 FARKAS LEMMA

Farkas Lemma [F2] (see also [F6] and [Z3]) forms the basis for obtaining more readily applied necessary conditions for constrained local extrema.

LEMMA 2.1 [Farkas Lemma]
Let $\{v_k\}$, $k = 0, 1, 2, \ldots, m$, be a set of n×1 vectors. A necessary and sufficient condition that there exist nonnegative scalar values η_k, $k = 1, 2, \ldots, m$, such that

$$v_0 = \sum_{k=1}^{m} \eta_k v_k \qquad (2\text{-}72)$$

is that for every vector Δx such that $\Delta x' v_k \leq 0$, $k = 1, 2, \ldots, m$, it follows that $\Delta x' v_0 \leq 0$.

<u>Proof</u>. Let

$$A \triangleq \{\Delta x \mid \Delta x' v_k \leq 0, \quad k = 1, 2, \ldots, m\} \quad (2\text{-}73)$$

If $v_0 = \sum_{k=1}^{m} \eta_k v_k$ and $\eta_k \geq 0$ for all k, then for all $\Delta x \in A$,

$$\Delta x' v_0 = \Delta x' \sum_{k=1}^{m} \eta_k v_k = \sum_{k=1}^{m} \eta_k (\Delta x' v_k) \quad (2\text{-}74)$$

With $\Delta x \in A$, each $\Delta x' v_k$ in (2-74) is nonpositive, whereas each $\eta_k \geq 0$, and therefore

$$\Delta x' v_0 \leq 0$$

Converse: We are given that $\Delta x' v_0 \leq 0$ for all $\Delta x \in A$ and must show that

$$v_0 = \sum_{k=1}^{m} \eta_k v_k \quad (2\text{-}75)$$

for some η_k's ≥ 0. Let

2.9 CONDITIONS OF OPTIMALITY

$$C \triangleq \{y \in E^n |\ y = \sum_{k=1}^{m} \alpha_k v_k,\ \text{for all}\ \alpha_k \geq 0\} \quad (2\text{-}76)$$

The set C is a <u>convex polyhedral cone</u> in E^n, having properties

i) for any $y \in C$ and scalar $\lambda \geq 0$, $\lambda y \in C$; and

ii) for $y_1, y_2 \in C$, $(y_1 + y_2) \in C$.

We wish to show that $v_0 \in C$. Assume the contrary, i.e., assume $v_0 \notin C$. We shall see that this assumption leads to a contradiction of the given fact that $\Delta x' v_0 \leq 0$ for all $\Delta x \in A$. Let z be the projection of v_0 onto C such that $z + \gamma = v_0$ as is illustrated for the two-dimensional case in Figure 2-5. Note that

$$\gamma'(y - z) \leq 0 \quad \text{and} \quad \gamma' z = 0$$

Thus,

$$\gamma' y \leq 0 \quad (2\text{-}77)$$

for any $y \in C$, and in particular,

$$\gamma' v_k \leq 0,\ k = 1, 2, \ldots, m,$$

which implies that $\gamma \in A$. But

$$\gamma' v_0 = \gamma'(z + \gamma) = \gamma' \gamma > 0 \quad (2\text{-}78)$$

This contradicts the fact that $\Delta x' v_0 \leq 0$ for all $\Delta x \in A$. Hence, the assumption that $v_0 \notin C$ is invalid, and therefore

$$v_0 = \sum_{k=1}^{m} \eta_k v_k, \quad \eta_k\text{'s} \geq 0 \qquad (2\text{-}79)$$

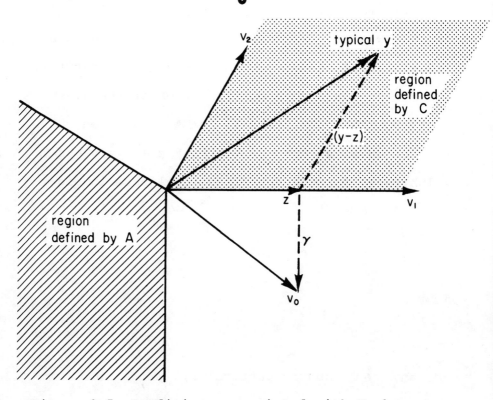

Figure 2-5 Conditions associated with Farkas Lemma.

2.10 CONDITIONS OF OPTIMALITY

2.10 EXISTENCE OF GENERALIZED LAGRANGE MULTIPLIERS: KUHN-TUCKER CONDITIONS

The following theorem is given for the case that S_2^* of (2-70) equals ϕ.

THEOREM 2.3 [Existence of Generalized Lagrange Multipliers]

If i) x^* satisfies constraints (2-2) and (2-3) and yields a constrained local maximum of $f(x)$,

ii) the NLP functions are of class C^1 in an open neighborhood of x^*, and

iii) $S_2^* = \phi$ (the normal case)

then there exist vectors α^*, β^* such that (x^*, α^*, β^*) satisfies the following conditions

$$p_i(x) = a_i, \quad i = 1, 2, \ldots, m_1 \qquad (2\text{-}80a)$$

$$q_j(x) \leq b_j, \quad j = 1, 2, \ldots, m_2 \qquad (2\text{-}80b)$$

$$\beta_j[b_j - q_j(x)] = 0, \quad j = 1, 2, \ldots, m_2 \qquad (2\text{-}80c)$$

$$\beta_j \geq 0, \quad j = 1, 2, \ldots, m_2 \qquad (2\text{-}80d)$$

and

$$\nabla L = 0 \qquad (2\text{-}80e)$$

Proof. For all i write $\{p_i(x^*) = a_i\}$ as two inequalities $\{p_i(x^*) \leq a_i\}$ and $\{p_i(x^*) \geq a_i\}$. Assumption iii that $S_2^* = \phi$ implies that $\Delta x = x - x^*$ belongs to S_1^*. Then by Lemma 2.1 there exist nonnegative scalar values u_i^*, $i = 1, 2, \ldots, m_1$, v_i^*, $i = 1, 2, \ldots, m_1$, and β_j^*, $j \in S_a$, such that

$$\nabla f^* = \sum_{i=1}^{m_1} (u_i^* - v_i^*) \nabla p_i^* + \sum_{j \in S_a} \beta_j^* \nabla q_j^* \quad (2\text{-}81)$$

Let $\alpha_i^* = (u_i^* - v_i^*)$ for all i and $\beta_j^* = 0$ for all $j \notin S_a$. Then

$$\nabla f^* - \sum_{i=1}^{m_1} \alpha_i^* \nabla p_i^* - \sum_{j=1}^{m_2} \beta_j^* \nabla q_j^* = 0$$

which is just condition (2-80), $\nabla L^* = 0$, where $\beta_j^* \geq 0$ for $q_j^* = b_j$, and $\beta_j^* = 0$ for $q_j^* < b_j$. ●

Relations (2-80) are the <u>Kuhn-Tucker relations</u> that constitute a necessary condition that x^* be an optimal solution to the NLP in the normal case. Note that assumption iii implies that the conditions of Lemma 2.1 are satisfied by the set G at x^* where

2.10 CONDITIONS OF OPTIMALITY

$$G = \{\nabla f^*, \nabla p_i^* \ (i = 1, 2, \ldots, m_1),$$

$$-\nabla p_i^* \ (i = 1, 2, \ldots, m_1), \text{ and}$$

$$\nabla q_j^* \ (j \in S_a)\} \qquad (2\text{-}82)$$

To ensure the existence of finite multipliers when applying Theorem 2.3, we should determine if $S_2^* = \phi$. A useful criterion that can be used to test for $S_2^* = \phi$ is given in Theorem 2.4, which requires the following definitions.

DEFINITION 2.5 [Linear Independence]
The vectors $v_1, v_2, \ldots, v_m$ are said to be <u>linearly independent</u> if

$$c_1 v_1 + c_2 v_2 + \cdots + c_m v_m = 0 \qquad (2\text{-}83)$$

where $c_1, c_2, \ldots, c_m$ are constants, implies that $c_1 = c_2 = \cdots = c_m = 0$. Conversely, vectors $v_1, v_2, \ldots, v_m$ are said to be <u>linearly dependent</u> if and only if v_i can be expressed as a linear combination of v_j ($j = 1, 2, \ldots, m; \ j \neq i$).

DEFINITION 2.6 [Regular Point]

A point at which the gradients of the active constraints are linearly independent is called a <u>regular point</u>, and any constrained local maximum that occurs at a regular point is said to satisfy the <u>regularity condition</u>.

THEOREM 2.4 [Sufficient Condition that $S_2^* = \phi$]

A sufficient condition that $S_2^* = \phi$ (and therefore that condition iii of Theorem 2.3 is satisfied) is that the gradients ∇p_i, $i = 1, 2, \ldots, m_1$, and ∇q_j, $j \in S_a$, are linearly independent at a constrained local maximum point x^*; that is, $S_2^* = \phi$ if x^* is a regular point.

<u>Proof</u>. The feasible region is defined by $p_i(x) = a_i$ ($i = 1, 2, \ldots, m_1$) and $q_j(x) \leq b_j$ ($j = 1, 2, \ldots, m_2$). The point x^* is in the feasible region and yields a constrained local maximum of $f(x)$. For the moment, let us replace the active inequality constraints at x^* by corresponding equality constraints to form a potentially smaller feasible region. Clearly, $f(x^*)$ is also a constrained local maximum of $f(x)$ in this new region defined by

2.10 CONDITIONS OF OPTIMALITY

$p_i(x) = a_i$ (all i) and $q_j(x) = b_j$ (all $j \in S_a$). For this equality constraint case, the linearly independent constraints define a tangent plane at x^*; that is, Δx is in the tangent plane if it is in the set $\{\Delta x \mid \Delta x' \nabla p_i^* = 0$ (all i), and $\Delta x' \nabla q_j^* = 0$ (all $j \in S_a$)$\}$. The gradient ∇f^* must be normal to this tangent plane because f^* is a constrained maximum. Therefore, ∇f^* must be expressible as a linear combination of the linearly independent ∇p_i^*'s and ∇q_j^*'s. Thus,

$$\nabla f^* = \sum_{i=1}^{m_1} \alpha_i \nabla p_i^* + \sum_{j \in S_a} \beta_j \nabla q_j^* \qquad (2\text{-}84)$$

Our goal at this point is to show that β_j's in (2-84) must satisfy $\beta_j \geq 0$: we can then invoke Lemma 2.1 (accounting for the original equality constraints in the same way as in the proof of Theorem 2.3) giving the equivalence between equation (2-84) and the desired result that $\Delta x' \nabla f^* \leq 0$ for every Δx which satisfies both $\Delta x' \nabla p_i^* = 0$ (all i) and $\Delta x' \nabla q_j^* \leq 0$ (all $j \in S_a$). The later statement is the same as the statement that $S_2^* = \phi$.

Assume the contrary, that is, assume

$$\beta_k < 0, \text{ for some } k \in S_a \tag{2-85}$$

Let V be the following tangent plane at x^*.

$$V = \{\Delta x \mid \Delta x'\nabla p_i^* = 0 \text{ (all i)},$$

$$\text{and } \Delta x'\nabla q_j^* = 0$$

$$\text{(all } j \in S_a \text{ except } j \neq k)\} \tag{2-86}$$

Because of the linear independence of the active constraint gradients at x^*, we can find a vector $z \in V$ such that $z'\nabla q_k^* < 0$. We use this z vector to form a smooth parametric representation $x(t)$ of x with $x(0) = x^*$ and $dx(0)/dt = z$. We have then

$$\left.\frac{d}{dt} q_k(x(t))\right|_{t=0} = z'\nabla q_k^* < 0 \tag{2-87}$$

which points into the feasible set. Also

$$\left.\frac{d}{dt} f(x(t))\right|_{t=0} = z'\nabla f^* \tag{2-88}$$

Consider the result of multiplying both sides of (2-84) by z'. After accounting for conditions

2.10 CONDITIONS OF OPTIMALITY

(2-85), (2-87), and $z \in V$ of (2-86), we have

$$z'\nabla f^* = \beta_k z'\nabla q_k^* > 0 \qquad (2\text{-}89)$$

But (2-88) and (2-89) give that $df(x(0))/dt > 0$ which contradicts the assumption that f^* is a constrained maximum. Thus, β_k must satisfy $\beta_k \geq 0$. ●

Theorem 2.3 is a first-order characterization of local maxima in that it involves first-order differentiability of the problem functions. It therefore does not take into account the curvature of the functions if they are nonlinear.

2.11 SUFFICIENT CONDITIONS FOR CONSTRAINED MAXIMA

In Theorem 2.2, we developed second-order sufficient conditions for the unconstrained problem. By analogy with the unconstrained problem, we could hypothesize that $\nabla^2 L^*$ be negative definite on the tangent hyperplane H formed by the equality constraints and active inequality constraints at x^*. In most situations this is sufficient. But in the case of <u>degenerate inequality</u>

constraints (that is, constraints with the property that $q_j(x^*) = b_j$ and $\beta_j^* = 0$), we must require $\nabla^2 L^*$ to be negative definite on a hyperplane larger than H.

Sufficient conditions based on the Kuhn-Tucker conditions of Theorem 2.3 are given in the following theorem.

THEOREM 2.5 [A Set of Sufficient Conditions for a Constrained Local Maximum]

If   i) x^* satisfies (2-2) and (2-3),

ii) the NLP functions are of class C^2 in an open neighborhood of x^*,

iii) the gradients ∇p_i, $i = 1, 2, \ldots, m_1$, and ∇q_j, $j \in S_a$, are linearly independent at x^* (i.e., x^* is a regular point), and

iv) the set C_a is defined by $C_a \triangleq j | \beta_j > 0\}$

then sufficient conditions that $f(x^*)$ be a strict constrained local maximum are that there exist vectors α^* and β^* such that (x^*, α^*, β^*) satisfies conditions (2-80) and that for every nonzero vector Δx satisfying $\Delta x' \nabla p_i^* = 0$, $i = 1, 2, \ldots, m_1$, $\Delta x' \nabla q_j^* = 0$ for $j \in C_a$, and $\Delta x' \nabla q_k^* \leq 0$ for $k \in S_a - C_a$, it follows that

2.11 CONDITIONS OF OPTIMALITY

$$\Delta x' \nabla^2 L^* \Delta x < 0 \qquad (2\text{-}90)$$

Proof. We have

$$\Delta f \triangleq f(x) - f^*$$
$$= \Delta x' \nabla f^* + \tfrac{1}{2} \Delta x' \nabla^2 f^* \Delta x + \cdots \qquad (2\text{-}91)$$

$$\Delta p_i \triangleq p_i(x) - p_i^*$$
$$= \Delta x' \nabla p_i^* + \tfrac{1}{2} \Delta x' \nabla^2 p_i^* \Delta x + \cdots \qquad (2\text{-}92)$$

and

$$\Delta q_j \triangleq q_j(x) - q_j^*$$
$$= \Delta x' \nabla q_j^* + \tfrac{1}{2} \Delta x' \nabla^2 q_j^* \Delta x + \cdots \qquad (2\text{-}93)$$

where $x = x^* + \Delta x$. We multiply (2-92) and (2-93) by α_i^* and β_j^*, respectively, and subtract the results from (2-91) to obtain

$$\Delta f - \sum_{i=1}^{m_1} \alpha_i^* \Delta p_i - \sum_{j \in C_a} \beta_j^* \Delta q_j$$
$$= \Delta x' \nabla L^* + \tfrac{1}{2} \Delta x' \nabla^2 L^* \Delta x + \cdots \qquad (2\text{-}94)$$

Condition (2-80e) gives $\nabla L^* = 0$. Also, if we consider only allowed Δx, Δp_i is 0 for all i, and (2-94) reduces to

$$\Delta f = \sum_{j \in C_a} \beta_j^* \Delta q_j + \tfrac{1}{2}\Delta x' \nabla^2 L^* \Delta x + \cdots \qquad (2-95)$$

Consider an allowed Δx sufficiently small in magnitude for which $\Delta q_j < 0$ or equivalently $\Delta x' \nabla q_j^* < 0$. Such a first-order term would dominate the right-hand side of (2-95) and guarantee $\Delta f < 0$. Thus, we need only consider those Δx for which $\Delta x' \nabla q_j^* = 0$ for $j \in C_a$. This, along with condition iii, enables us to view the q_j's for $j \in C_a$ in the same manner as equality constraints, insofar as sufficiency is concerned, in which case

$$\Delta f = \tfrac{1}{2}\Delta x' \nabla^2 L^* \Delta x + O(\|\Delta x\|^3) \qquad (2-96)$$

Thus, the incremental change Δf will be strictly less than zero for all allowed Δx's that are sufficiently small in magnitude if $\tfrac{1}{2}\Delta x' \nabla^2 L^* \Delta x < 0$ for $\Delta x \in \{\Delta x \mid \Delta x' \nabla p_i^* = 0 \text{ (all } i\text{)}, \Delta x' \nabla q_j^* = 0 \text{ (all } j \in C_a\text{)}, \text{ and } \Delta x' \nabla q_k^* \leq 0 \text{ (all } k \in S_a - C_a\text{)}\}$.

•

If (2-90) is replaced by $\Delta x' \nabla^2 L^* \Delta x \leq 0$, the resulting conditions of Theorem 2.5 are second-order necessary conditions for a regular point x^* to yield a constrained local maximum.

2.11 CONDITIONS OF OPTIMALITY

EXAMPLE 2.7 Consider the following problem:

$$\text{maximize } f(x) = x_2 \quad (2\text{-}97)$$

subject to

$$p_1(x) = x_1^2 + x_2^2 = 1 \quad (2\text{-}98a)$$

$$q_1(x) = 2x_2 - x_1 \leq 1 \quad (2\text{-}98b)$$

and $x_1 \geq 0$, or

$$q_2(x) = -x_1 \leq 0 \quad (2\text{-}98c)$$

The Lagrangian is

$$L = x_2 + \alpha_1(1 - x_1^2 - x_2^2)$$
$$+ \beta_1(1 + x_1 - 2x_2) + \beta_2 x_1 \quad (2\text{-}99)$$

for which the gradient is

$$\nabla L = \begin{bmatrix} -2\alpha_1 x_1 + \beta_1 + \beta_2 \\ \\ 1 - 2\alpha_1 x_2 - 2\beta_1 \end{bmatrix} \quad (2\text{-}100)$$

To satisfy the first-order Kuhn-Tucker conditions, $x^* = (3/5, 4/5)$ at which $f(x^*) = 4/5$, $\alpha_1^* = 1/4$, and $\beta^* = (3/10, 0)$.

Note that the active constraints at x^* are (2-98a) and (2-98b), and that S_a equals C_a because $\beta_1^* > 0$. The gradients of interest are $\nabla p_1^* = [1.2\ \ 1.6]'$ and $\nabla q_1^* = [-1\ \ 2]'$ which are linearly independent. Thus, for sufficiency, we need only check to see if $\Delta x' \nabla^2 L^* \Delta x < 0$ for all nonzero Δx that satisfy both $\Delta x' \nabla p_1^* = 0$ and $\Delta x' \nabla q_1^* = 0$, but no such Δx exists. Here, condition (2-90) of Theorem 2.5 is satisfied in a vacuous sense, and second-order sufficient conditions are satisfied.

For completeness, consider $\nabla^2 L$,

$$\nabla^2 L = \begin{bmatrix} -2\alpha_1 & 0 \\ 0 & -2\alpha_1 \end{bmatrix} \qquad (2\text{-}101)$$

At (x^*, α^*, β^*), $\Delta x' \nabla^2 L^* \Delta x = -\tfrac{1}{2}(\Delta x_1^2 + \Delta x_2^2) < 0$ for any nonzero Δx. However, this last result is less important than the sufficiency result of the preceding paragraph.

EXAMPLE 2.8 Reconsider the parcel problem (Example 2.4). The Hessian $\nabla^2 L^*$ is

2.11 CONDITIONS OF OPTIMALITY

$$\nabla^2 L^* = \begin{bmatrix} 0 & 1 & 1 \\ 1 & 0 & 1 \\ 1 & 1 & 0 \end{bmatrix} \qquad (2\text{-}102)$$

Note that $-\nabla^2 L^*$ is not a positive-definite matrix. For any Δx,

$$\Delta x' \nabla^2 L^* \Delta x = [\Delta x_1 \ \Delta x_2 \ \Delta x_3] \begin{bmatrix} 0 & 1 & 1 \\ 1 & 0 & 1 \\ 1 & 1 & 0 \end{bmatrix} \begin{bmatrix} \Delta x_1 \\ \Delta x_2 \\ \Delta x_3 \end{bmatrix}$$

$$= 2\Delta x_1 (\Delta x_2 + \Delta x_3) + 2\Delta x_2 \Delta x_3 \qquad (2\text{-}103)$$

However, with $\beta_2 = 0.09$, we restrict our attention to Δx's which satisfy $\Delta x' \nabla q_2^* = 0$ where $q_2^* = 1.8$ is the only active constraint at x^*. Thus,

$$[\Delta x_1 \ \Delta x_2 \ \Delta x_3] \begin{bmatrix} 1 \\ 2 \\ 2 \end{bmatrix} = \Delta x_1 + 2\Delta x_2 + 2\Delta x_3 = 0 \quad (2\text{-}104)$$

or

$$\Delta x_1 = -2\Delta x_2 - 2\Delta x_3 \qquad (2\text{-}105)$$

which is used in (2-103) to obtain

$$\Delta x' \nabla^2 L^* \Delta x = -4(\Delta x_2 + \Delta x_3)^2 + 2\Delta x_2 \Delta x_3$$
$$= -4(\Delta x_2 + 0.75\Delta x_3)^2 - 1.75\Delta x_3^2$$
$$(2\text{-}106)$$

which is less than zero if $\Delta x \neq 0$, and therefore second-order sufficient conditions are satisfied.

2.12 SUFFICIENT CONDITIONS AND THE LINEAR PROBLEM

For the linear problem, the constraints and the objective function have matrices of second derivatives with each entry identically equal to zero. Thus, (2-90) cannot hold at an optimal point x^*. However, consider the typical linear case in which both C_a equals S_a and there are exactly n active constraints, the gradients of which are linearly independent, at x^*. Here there are no nonzero Δx which satisfy the n equations of the set $\{\Delta x' \nabla p_i^* = 0$ for all i, and $\Delta x' \nabla q_j^* = 0$ for $j \in S_a\}$, with the result that the tangent hyperplane formed by the active constraints is empty. In this typical linear case, therefore, the solution point is unique and is given by an extreme

2.12 CONDITIONS OF OPTIMALITY

point of the convex polyhedron defined by the constraints.

For certain <u>atypical</u> linear cases (Problem 2.5c, for example), fewer than n independent constraints may be active at an optimal x^*, and the solution point x^* need not be unique. Even here, however, the convexity considerations of Chapter 1 ensure us that there is but one constrained local maximum value f^* of f (assuming f^* exists) and that f^* is the global solution to the problem.

EXAMPLE 2.9 Consider the following linear problem:

$$\text{maximize } f = \tfrac{1}{2}x_1 + x_2 + \tfrac{1}{2}x_3 + x_4 \qquad (2\text{-}107)$$

subject to

$$p_1 = x_1 + x_2 + x_3 - 2x_4 = 6$$

$$q_1 = x_1 + x_2 + x_3 + x_4 \leq 10$$

$$q_2 = 0.2x_1 + 0.5x_2 + x_3 + 2x_4 \leq 10 \qquad (2\text{-}108)$$

$$q_3 = 2x_1 + x_2 + 0.5x_3 + 0.2x_4 \leq 10$$

and

$$q_{3+i} = -x_i \leq 0, \quad i = 1, 2, 3, 4$$

The optimum point is $x^* = (0, 26/3, 0, 4/3)$, $f(x^*) = 10$, $\alpha_1^* = 0$, and $\beta^* = (1, 0, 0, \frac{1}{2}, 0, \frac{1}{2}, 0)$. The gradients of the active constraints at x^*, i.e., ∇p_1^*, ∇q_1^*, ∇q_4^*, and ∇q_6^* are linearly independent as the n×n matrix formed by $[\nabla p_1^*, \nabla q_1^*, \nabla q_4^*, \nabla q_6^*]$ can be reduced through elementary matrix manipulations, to the n×n identity matrix I. Thus, x^* is the unique solution.

2.13 SUMMARY

When the functions of an NLP are of class C^1 over an extended feasible region, Kuhn-Tucker first-order conditions are satisfied at all constrained local maxima, provided they fall within the class of normal cases. An in-depth study of the difference between normal and abnormal cases requires constraint qualification concepts [A1, G4, F6]; the subtleties of these concepts are avoided in this chapter. We found, however, that a sufficient condition for normality at a point x^* is that the active constraints satisfy a regularity condition at x^*, namely, that the gradients of the active constraints be linearly independent

2.13 CONDITIONS OF OPTIMALITY

at x^*. While this regularity condition is usually satisfied in practice, it is not a necessary condition for normality (see Problem 2.10).

In developing second-order conditions, we restricted our attention to those points of constrained local maxima at which the regularity condition holds. For second-order sufficient conditions, we required that first-order conditions be satisfied, and that the Hessian $\nabla^2 L^*$ of the Lagrangian L satisfy $\Delta x' \nabla^2 L^* \Delta x < 0$ for a particular set of Δx's that depend on the gradients of the active constraint functions. Thus, even though $\nabla L^* = 0$ at a constrained local maximum point, L^* need not be an unconstrained local maximum of L. In Chapter 4, we expand the above development by considering an augmented Lagrangian L_a that does exhibit an unconstrained local maximum with respect to x at (x^*, α^*, β^*), subject to a set of general conditions.

PROBLEMS

2.1 Apply local optimality conditions to the following functions:

a. $ax^2 + bx + c$

b. $ax^3 + bx^2 + cx + d$

c. The function of Figure 1-7.

d. $e^{-x} - e^{-2x}$

e. $x_2^4 - 2x_1 x_2^2 + 2x_1^2 + 2x_1 x_2 + x_2^2 - 4x_1 - 4x_2 + 4$

[examine $x_1 = x_2 = 1$ and $(x_1, x_2) = (4, -2)$]

2.2 Any unconstrained local maximum must satisfy the second-order necessary condition that is stated following Theorem 2.2. Supply a proof for this second-order necessary condition.

2.3 Given $\nabla f^* = 0$, what do the following $\nabla^2 f^*$ matrices suggest in regard to unconstrained local optima of f?

a. $\begin{bmatrix} 1 & -1 \\ -1 & 2 \end{bmatrix}$

b. $\begin{bmatrix} 1 & 1 \\ 1 & -2 \end{bmatrix}$

c. $\begin{bmatrix} -1 & 1 \\ 1 & -2 \end{bmatrix}$

d. $\begin{bmatrix} 1 & 1 \\ 1 & 1 \end{bmatrix}$

2.4 Measurements on a process supply the matrix equation $Ax - b = \varepsilon$, where A is m×n with m > n,

PROBLEMS CONDITIONS OF OPTIMALITY

b and ε are m×1, and x is n×1. A and b contain known real values, ε is a measurement error vector, and x is to be selected to minimize the squared error $\varepsilon'\varepsilon$. Apply first- and second-order optimality conditions and determine the unconstrained minimum of $\varepsilon'\varepsilon$ with respect to x.

2.5 Constraints are $x_1 + x_2 \leq 3$, $0 \leq x_1 \leq 2$, and $0 \leq x_2 \leq 2$. To be maximized is f where:
a) $f = 2x_1 + x_2$; b) $f = x_1 + 2x_2$; and c) $f = x_1 + x_2$. Portray first-order conditions graphically (see Figure 2-2).

2.6 To be minimized is the square of the distance from the origin of the (x_1, x_2) plane to the curve characterized by $x_1^2 - cx_2 = c$ where c is a real constant.

   a. Obtain a solution point by using substitution followed by unconstrained minimization with respect to one variable. Substitute $c(1 + x_2)$ for x_1^2 in the distance-squared function.

   b. Obtain the solution by using the Lagrange multiplier approach.

   c. For what values of the constant c is the solution of part a invalid? Explain.

2.7 Give first-order optimality conditions for the production problem 1.1. If all $x_i > 0$ in the optimal solution, show that the optimality conditions support the following economic principle: with p = 10,000 representing the equality constraint, the <u>marginal productivities</u> $(\partial p/\partial x_i)^*$, i = 1, 2, 3, are equal. That is, at the optimum, an incremental increase in the units produced at one plant will cost the same amount as a corresponding incremental increase in the units produced at any other plant.

2.8 Give the economic generalization of Problem 2.7 for the case that cost $C = x_1 + x_2 + x_3$ is replaced by a weighted measure $r_1 x_1 + r_2 x_2 + r_3 x_3$, the r_i's being constants.

2.9 Given to be maximized is $f = x_2$ subject to satisfying $x_1^2 + x_2^2 + x_3^2 = 1$ and $2x_2 - x_1 \leq 1$. Verify that first-order conditions are satisfied at points $x = (0.6, 0.8, 0)$ and $x = (-1, 0, 0)$. What do second-order conditions tell us in regard to the above points?

2.10 Maximize $f = x_1 + 2x_2$ subject to $x_1 \leq 2$, $x_2 \leq 2$, and $(x_1 - 1)^2 + (x_2 - 1)^2 \leq 2$.

    a. Depict the problem graphically and find the constrained maximum point.

    b. Does this problem satisfy the regularity condition (see Definition 2.6)?

    c. Is condition iii of Theorem 2.3 satisfied at the optimum?

    d. Establish the conditions that the multipliers must satisfy. If $\beta_2 = 1$ is assigned, what values of β_1 and β_3 must apply?

2.11 We are given the problem of maximizing $f(x)$ subject to $p_1(x) = c_1$ and $p_2(x) = c_2$. We observe that the two constraints are equivalent to one constraint as follows: $(p_1(x) - c_1)^2 + (p_2(x) - c_2)^2 = 0$. It is suggested that we solve the problem by using just one multiplier and the Lagrangian

PROBLEMS CONDITIONS OF OPTIMALITY

$f(x) - \alpha_1[(p_1(x) - c_1)^2 + (p_2(x) - c_2)^2]$. What is the falacy in this suggestion?

2.12 Solve the architect's problem, Problem 1.2.

2.13 For the electrical power source problem, Problem 1.7, a maximum of $f = x_1^2 x_2$ is desired, subject to satisfying $10 = x_1(10 + x_1 + x_2)$. Form the appropriate Lagrangian and solve for optimal values of x_1 and x_2.

2.14 Consider the inequality constrained problem having the Lagrangian $L_a = f - \sum_{i=1}^{m} \beta_i(q_i - b_i)$.
If $f(x)$ is concave and all q_i's are convex, what conclusions can be drawn in regard to L_a and the solution of the problem?

2.15 Verify that the gradients of the active constraints of Example 2.9 are linearly independent.

2.16 Are the gradients of the active constraints of Problem 1.9 linearly independent for $\ell = 2$ and $m = 3$? Explain why this is so.

2.17 Derive the Kuhn-Tucker conditions for the following NLP formulation.

minimize $f(x)$

subject to

$p_i(x) = a_i$, $\quad i = 1, 2, \ldots, m_1 < n$

$q_j(x) \geq b_j$, $\quad j = 1, 2, \ldots, m_2$

108   MATH PROGRAMMING VIA AUGMENTED LAGRANGIANS

2.18   If x_1 and x_2 are related by $2x_1 + x_2 = 2$, find the points on the ellipsoid $2x_1^2 + x_2^2 + 4x_3^2 = 4$ which are, respectively, the farthest and closest from the origin.

2.19   Find the extrema of the function $f(x) = x_1^2 + x_2^2 - 4x_1 x_2$ on the circle $x_1^2 + x_2^2 = a$.

2.20   What are the dimensions of a parallelepiped with diagonal d that will maximize its volume?

2.21   Given the following system of equations

$$2x_1 - 5x_2 + x_3 = 1$$
$$x_1 + 2x_2 - x_3 = 2$$
$$x_1 - 2x_2 + 2x_3 = 4$$

give a quadratic function that has a minimum point at the solution of the above system. Show that the two problems are equivalent.

2.22   Solve Problem 1.8 for (x^*, α^*, β^*).

2.23   Given the functions

$$f(x) = -3(x_1 - 2)^2 - (x_2 - 3)^2 - 2(x_3 - 1)^2$$
$$q_1(x) = 2x_1 + x_2 + 3x_3$$
$$q_2(x) = x_1^2 + 2x_2^2 + 3x_3^2$$

solve the following using Lagrange multipliers.

  a. max $f(x)$, $q_1(x) \geq 6$, $q_2(x) \geq 30$
  b. max $f(x)$, $q_1(x) \geq 6$, $q_2(x) \leq 30$

PROBLEMS CONDITIONS OF OPTIMALITY   109

c. max $f(x)$, $q_1(x) \leq 6$, $q_2(x) \leq 30$

d. min $(-f(x))$, $q_1(x) \geq 6$, $q_2(x) \leq 30$

e. min $(-f(x))$, $q_1(x) \leq 6$, $q_2(x) \leq 30$

2.24 Consider the following problem from [C7].

$$\text{Minimize } f = x_1^2 + x_2^2$$

subject to

$$(x_1 - 1)^3 - x_2^2 = 0$$

Solve using the Lagrangian and verify graphically. What can be said about the behavior of this problem?

2.25 Consider the following constraints in E^3:

$$(x_1 - 1)^2 + x_2 - (x_3 - 2)^3 \leq 0,$$

$$0 \leq x_1, \quad 0 \leq x_2, \quad \text{and} \quad 0 \leq x_3 \leq 2$$

Show that the point (1, 0, 2) is a feasible point. Is it also a regular point?

3
SENSITIVITY

3.1 INTRODUCTION

A practical question that often arises in solving either a mathematical or a real-world problem is how much the optimum changes when changes are made in the constraints or when changes are made in the optimal parameter values or system elements. Sensitivity analysis allows one to examine the variations in an objective or cost function when such changes are made. Physical systems cannot be constructed or operated exactly as in theory because of the inherent limitations on man's capabilities and the limitations caused by a real-world environment. But these systems may be constructed and operated satisfactorily if the elements that influence system operation maintain certain tolerance limits. Tolerance specifications

3.1 SENSITIVITY   111

are an important part of a complete solution to a
practical problem, and to specify them realistically one must know how changes in the system elements influence the characteristics of the system
or problem under consideration. Sensitivity analysis provides the means of obtaining such information.

3.2 MICROSCOPIC SENSITIVITY

The fractional change in a given characteristic, which is a function of system parameters,
can be estimated on the basis of fractional changes
in the parameters. When these fractional changes
are required to remain within specified bounded
regions or tolerance limits, worst-case sensitivities of system characteristics may be estimated by
directly incorporating tolerance limits in the
performance measure and constraint equations. For
example, system parameters are set to tolerance
extremes which are most likely to cause constraint
violation and are least favorable to optimal performance. The performance measure is then optimized under these worst-case conditions. A

similar analysis could be performed where the tolerance limits are at the opposite extremes and the performance measure is optimized under best-case conditions yielding best-case sensitivities. Based on the above type of analysis, a designer can choose design-center values for system parameters. Microscopic sensitivity can then be analyzed.

Microscopic (incremental) sensitivity measures are usually based on a truncated Taylor series, since the incremental sensitivity of a system characteristic is usually applicable to small regions within the tolerance extremes about some design center. Let us consider the variation of the performance measure $f(x^*)$ where x^* is the design-center value of the system parameter vector x. If we assume that $f(x)$ is of class C^2, we can expand $f(x)$ about x^* in the truncated Taylor series

$$f(x) \cong f(x^*) + (x - x^*)'\nabla f(x^*)$$
$$+ \tfrac{1}{2}(x - x^*)'\nabla^2 f(x^*)(x - x^*) \quad (3\text{-}1)$$

Define Γ and Λ as follows:

$$\Gamma \triangleq \left[\frac{x_1 - x_1^*}{x_1^*}, \frac{x_2 - x_2^*}{x_2^*}, \ldots, \frac{x_n - x_n^*}{x_n^*} \right]' \quad (3\text{-}2)$$

3.2 SENSITIVITY

and

$$\Lambda \triangleq \text{diag}(x_1^*, x_2^*, \ldots, x_n^*) \tag{3-3}$$

where Λ is an n×n diagonal matrix. With these identities, (3-1) can be rearranged as in [P3] to obtain

$$\frac{f(x) - f(x^*)}{f(x^*)} = \Gamma' \frac{\Lambda \nabla f(x^*)}{f(x^*)} + \Gamma' \frac{\Lambda \nabla^2 f(x^*) \Lambda}{2f(x^*)} \Gamma$$

$$= \Gamma' S_x^f + \Gamma' Sq_x^f \Gamma \tag{3-4}$$

where the column sensitivity matrix S_x^f is defined by

$$S_x^f \triangleq \frac{\Lambda \nabla f(x^*)}{f(x^*)} \tag{3-5}$$

and the square sensitivity matrix Sq_x^f is defined by

$$Sq_x^f \triangleq \frac{\Lambda \nabla^2 f(x^*) \Lambda}{2f(x^*)} \tag{3-6}$$

Both $f(x)$ and x^* are assumed known so that S_x^f and Sq_x^f can be evaluated. If x^* happens to be an unconstrained local maximum (or minimum),

$S_x^f = 0$. Equation (3-4) yields the fractional change in f(x) which results from the fractional changes in x as specified in the Γ matrix. If $f(x^*) = 0$ or if any $x_i^* = 0$, however, equations (3-4), (3-5), and (3-6) should not be used -- in such cases, equation (3-1) should be considered directly with sensitivity measured in magnitude of change rather than in fractional change.

EXAMPLE 3.1 For the area-volume problem (Example 2.2), it can be shown that $S_x^f = [2 \quad 1]'$ and

$$Sq_x^f = \begin{bmatrix} 1 & 1 \\ 1 & 0 \end{bmatrix}$$

which are used in (3-4) to obtain

$$\frac{\Delta f}{f^*} = \frac{\Delta x_1}{x_1^*}(2 + \frac{\Delta x_1}{x_1^*} + 2 \frac{\Delta x_2}{x_2^*}) + \frac{\Delta x_2}{x_2^*} \qquad (3-7)$$

where $\Delta f = f(x) - f(x^*)$, $\Delta x_1 = x_1 - x_1^*$, and $\Delta x_2 = x_2 - x_2^*$. With f being proportional to the volume of the tank, (3-7) gives that a 10% increase in radius x_1 from x_1^* produces approximately

3.2 SENSITIVITY

twice as much increase in volume as a corresponding 10% increase in height x_2. The area of the tank, however, also increases in the same general way (Problem 3.1).

3.3 MACROSCOPIC SENSITIVITY

Macroscopic sensitivity of a system characteristic is sensitivity with respect to large parameter changes. For example, if a range of values of a particular system parameter gives rise to a range of values for a system characteristic of interest, a design-center value may be selected so that the system characteristic is least sensitive to changes in the parameter. In many applications, tolerance limits ℓ_i are known for design-center values x_i^* such that, in matrix form,

$$-\ell \leq \Gamma \leq \ell \qquad (3-8)$$

where $\ell = [\ell_1 \; \ell_2 \; \cdots \; \ell_n]'$ and Γ is defined in (3-2).

The question that often arises then is: "What is the maximum of the absolute value of $[f(x) - f(x^*)]/f(x^*)$ in (3-4) with respect to allowed

116   MATH PROGRAMMING VIA AUGMENTED LAGRANGIANS

variations of Γ as in (3-8)?" This question is posed such that it can be put into the format of the NLP and can be answered by solving the NLP with search techniques such as presented in this text. In general, the application of search techniques results in the accumulation of large amounts of macroscopic information.

EXAMPLE 3.2   For the area-volume problem (Example 2.2), consider fractional tolerances $\pm \ell_1$ for radius and $\pm \ell_2$ for height. Expression (3-8) is equivalent in this case to

$$(1 - \ell_i) x_i^* \leq x_i \leq (1 + \ell_i) x_i^*, \quad i = 1, 2 \quad (3-9)$$

The volume of the tank is proportional to $f(x)$,

$$f(x) = x_1^2 x_2 \quad (3-10)$$

The smallest volume, caused by tolerance deviations from x^*, is obtained by minimizing $f(x)$ subject to (3-9). Here, it is clear that this worst-case volume is $(1 - \ell_1)^2 (1 - \ell_2) (x_1^*)^2 x_2^*$.

3.3 SENSITIVITY

The area of the tank is proportional to $p(x)$,

$$p(x) = x_1 x_2 + 0.5 x_1^2 \qquad (3-11)$$

The largest area, caused by tolerance deviations from x^*, is obtained by maximizing $p(x)$ subject to (3-9), with the result being $(1 + \ell_1)(1 + \ell_2) x_1^* x_2^* + 0.5(1 + \ell_1)^2 (x_1^*)^2$.

3.4 SENSITIVITY AND LAGRANGE MULTIPLIERS

Sensitivity in the broad sense of the term may be applied to changes in constraint conditions as well as to changes in the parameters. The Lagrange multipliers that satisfy the Kuhn-Tucker conditions are frequently termed <u>shadow prices</u> [B3, V3] or <u>attribute costs</u> because they have the property of giving an attributed cost to a constraint. Consider the general problem

$$\max\ f(x) \qquad (3-12)$$

subject to

$$p_i(x) = a_i, \quad i = 1, 2, \ldots, m_1 < n \qquad (3-13)$$

and

$$q_j(x) \leq b_j, \quad j = 1, 2, \ldots, m_2 \qquad (3\text{-}14)$$

As a general example, $f(x)$ may be viewed as a profit function for some production process; the constraints of (3-13) may be machine operation and manpower costs needed to produce a line of products from given resources while the constraints of (3-14) may represent the availability or limitation on the total amount of the jth resource. By some investment or reduction in resources, or by effecting a change in personnel or production policy, the manager confronted with the above problem could obtain displacements Δa_i, $i = 1, 2, \ldots, m_1$, in (3-13) and displacements Δb_j, $j = 1, 2, \ldots, m_2$, in (3-14). Assuming that by shifting the right-hand sides of the constraints (3-13) and (3-14) by small amounts Δa_j and Δb_j, and assuming that there are associated savings or costs π_i and ξ_j per unit of change, investment costs or savings $\pi_i \Delta a_i$ and $\xi_j \Delta b_j$ result. The following question than arises: "Is the investment or policy change profitable or not?"

3.4 SENSITIVITY

Of primary interest in answering this question is the local behavior of $f(x^* + \Delta x)$ as Δa_i and Δb_j are varied. We assume that (x^*, α^*, β^*) satisfies the Kuhn-Tucker conditions (2-80), and that the Δa_i's, Δb_j's, and Δx satisfy

$$p_i(x^* + \Delta x) = a_i + \Delta a_i, \quad \text{all } i \qquad (3\text{-}15a)$$

$$q_j(x^* + \Delta x) = b_j + \Delta b_j, \quad j \in S_a \qquad (3\text{-}15b)$$

and

$$q_j(x^* + \Delta x) < b_j + \Delta b_j, \quad j \notin S_a \qquad (3\text{-}15c)$$

We can expand f, p_i, and q_j as follows:

$$f(x^* + \Delta x) = f(x^*) + \Delta x' \nabla f^* + \cdots \qquad (3\text{-}16a)$$

$$p_i(x^* + \Delta x) = p_i(x^*) + \Delta x' \nabla p_i^* + \cdots \qquad (3\text{-}16b)$$

and

$$q_j(x^* + \Delta x) = q_j(x^*) + \Delta x' \nabla q_j^* + \cdots \qquad (3\text{-}16c)$$

where the terms represented by '$+ \cdots$' are of order $\|\Delta x\|^k$ with $k \geq 2$. Rearranging (3-16) using (3-15) we have

$$\Delta f = f(x^* + \Delta x) - f(x^*) = \Delta x' \nabla f^* + \cdots \qquad (3\text{-}17a)$$

$$\Delta a_i = p_i(x^* + \Delta x) - p_i(x^*) = \Delta x' \nabla p_i^* + \cdots \quad (3\text{-}17b)$$

and

$$\Delta b_j = q_j(x^* + \Delta x) - q_j(x^*) = \Delta x' \nabla q_j^* + \cdots \quad (3\text{-}17c)$$

If we multiply (3-17b) by α_i^* for each i and sum, we obtain

$$0 = \sum_{i=1}^{m_1} \alpha_i^* \Delta a_i - \Delta x' \sum_{i=1}^{m_1} \alpha_i^* \nabla p_i^* - \cdots \quad (3\text{-}18)$$

Similarly, (3-17c) can be multiplied by β_j^* for $j \in S_a$ and summed in the same way, with the result that

$$0 = \sum_{j \in S_a} \beta_j^* \Delta b_j - \Delta x' \sum_{j \in S_a} \beta_j^* \nabla q_j^* - \cdots \quad (3\text{-}19)$$

Now, (3-17a), (3-18), and (3-19) can be added to give

$$\Delta f = \Delta x' [\nabla f^* - \sum_{i=1}^{m_1} \alpha_i^* \nabla p_i^* - \sum_{j \in S_a} \beta_j^* \nabla q_j^*] + \nu \quad (3\text{-}20a)$$

and

3.4 SENSITIVITY

$$\nu \triangleq \sum_{i=1}^{m_1} \alpha_i^* \Delta a_i + \sum_{j \in S_a} \beta_j^* \Delta b_j + O(||\Delta x||^2) \quad (3\text{-}20b)$$

where $O(||\Delta x||^2)$ represents the combined effect of all terms of order $||\Delta x||^k$, $k \geq 2$. The term in brackets in (3-20a) is ∇L^* which must be zero for the Kuhn-Tucker conditions to be satisfied at (x^*, α^*, β^*). Thus, from (3-20) we obtain the desired result that

$$\Delta f = \sum_{i=1}^{m_1} \alpha_i^* \Delta a_i + \sum_{j \in S_a} \beta_j^* \Delta b_j + O(||\Delta x||^2) \quad (3\text{-}21)$$

If the Δa_i's and Δb_j's are small in magnitude so that the resulting Δx's in the neighborhood of x^* are small in magnitude, then $O(||\Delta x||^2)$ in (3-21) is negligible, and Δf varies as $\alpha_i^* \Delta a_i$ and $\beta_j^* \Delta b_j$. Because the multipliers β_j^*'s are required by the Kuhn-Tucker conditions to satisfy $\beta_j^* \geq 0$, $f(x^* + \Delta x)$ varies directly with Δb_j. The multipliers α_i^* are not restricted in sign, however, and care must be taken in interpreting the variation of $f(x^* + \Delta x)$ with Δa_i.

Thus, in regard to the management question, we see from (3-21) that when Δa_i and Δb_j are small, the answer is obtained by comparing π_i and ξ_j with the corresponding α_i^* and β_j^* values. The practical implication of this result is that we can easily obtain an indication of how the optimal value of the objective function changes as the resource or policy changes. In using a multiplier algorithm, the multipliers for the constraint functions are readily available at the solution of the problem, and the indications of favorable and unfavorable investments or policy changes are obtained without solving numerous NLP problems.

It must be noted, however, that large changes based on incremental indications may lead to results quite contrary to those expected, as the term $O(||\Delta x||^2)$ in (3-21) may dominate for large $||\Delta x||$. Also, the preceding analysis is restricted to cases where the active constraints remain invariant with respect to variations in a_i's and b_j's, i.e., S_a is fixed in the neighborhood of x^*. Large changes usually violate this assumption. Thus, before large-scale changes based on

3.4 SENSITIVITY

incremental indications are implemented, it is wise to solve the NLP under the new conditions.

EXAMPLE 3.3   Consider the application of (3-21) to Example 2.2. Let $\Delta a_1 = 0.1 a_1$ represent a 10% increase in area. It follows that $\Delta f \cong 0.1 a_1 \alpha_1^*$, and from (2-34), $\Delta f \cong 0.15 (\alpha_1^*)^3$. This result and $f^* = (\alpha_1^*)^3$ give $(\Delta f / f^*) \cong 0.15$, a 15% increase in volume f. In actuality, with the new a_1 in (2-30) equal to 1.1 times the old a_1, Example 2.2 can be solved again, giving a 15.37% increase in volume.

EXAMPLE 3.4   For the parcel problem of Example 2.4, equation (3-21) reduces to

$$\Delta f \cong \beta_2^* \Delta b_2, \quad \beta_2^* = 0.09 \qquad (3-22)$$

If b_2 is increased from 1.8 meters to 2 meters in (2-50), giving $\Delta b_2 = 0.2$, equation (3-22) gives $\Delta f \cong 0.018$ which represents an increase in parcel volume from 0.054 cubic meter to approximately 0.072 cubic meter. For an exact value, the original problem must be solved again, using 2 as the right-hand side of (2-50). The reader may verify that the new volume is actually $2/27 \cong 0.074$ cubic meter and the new x^* is $(2/3, 1/3, 1/3)$.

3.5 SUMMARY

The sensitivity analysis presented in this chapter gives a method for finding how the objective function changes when changes are made in the constraints or when changes are made in the optimal parameter values. Of most significance to the general NLP is the sensitivity information contained in the Lagrange multipliers that satisfy the Kuhn-Tucker conditions. In using a multiplier algorithm for solving an NLP, the multipliers are readily available at the solution. These multipliers have the property of giving an attributed cost to the problem constraints. With proper interpretation of the multipliers, indications of favorable investments or policy changes are obtained without solving numerous NLP problems.

PROBLEMS   SENSITIVITY

3.1 The function $x_1 x_2 + (x_1^2/2)$ represents the normalized area of the tank in Example 2.2. Find its column sensitivity matrix and its square sensitivity matrix at $x_1^* = x_2^* = 1$. Using equation (3-4), describe how a 10% increase in radius x_1 influences area; do the same for height x_2.

3.2 Derive a version of the sensitivity equation (3-21) that applies when f is minimized (-f is maximized) subject to constraints (3-13) and (3-14).

3.3 Consider a constraint of the form $g_j(x) \geq c_j$. This constraint can be incorporated in the set of constraints (3-14) by use of $-g_j(x) \leq -c_j$. Show that $-\beta_j \Delta c_j$ is the corresponding term that should be included in the sensitivity equation (3-21).

3.4 Give a detailed representation of the second-order terms that are omitted from (3-21).

3.5 Assume that the number 10,000 in Problem 1.1 is increased to 10,500. If $\alpha_1^* = 0.02$ for the 10,000 case and if f is in units of thousands of dollars, approximately how much additional cost will be incurred?

3.6 At the x = (0.6, 0.8, 0) solution point of Problem 2.9, estimate the increase in f that results when the right-hand side of the equality constraint is increased by 10% and that of the inequality constraint is increased by 5%.

3.7 From the results of Problem 2.22, determine which side of the triangle is most restrictive to the size of the inscribed circle. Using sensitivity information, answer the following:

a. What is the percentage change in the size of the circle if each side individually is allowed to move parallel to itself and away from the center by one unit?

b. What is the percentage change to be expected if all three sides are allowed to simultanteously move parallel to themselves and away from the center by one unit?

3.8 Check your results to Problem 3.7b by resolving the problem under the new constraints.

3.9 Interpret the variation of Δf to changes Δa_i and Δb_j in the NLP format given in Problem 2.17.

3.10 Perform a sensitivity study of any NLP for which you have obtained numerical results.

4
AN AUGMENTED LAGRANGIAN
AND
THE STRUCTURE FOR A MULTIPLIER ALGORITHM

4.1 INTRODUCTION

The previous chapters establish the central role played by the Lagrangian L and its multipliers in the normal solution to an NLP. The one not-so-nice feature of the Lagrangian, however, is the following: even though $\nabla L(x^*, \alpha^*, \beta^*) = 0$ at a constrained local maximum point x^*, $\nabla^2 L^*$ need not be negative definite, so that $L(x, \alpha^*, \beta^*)$ is not necessarily an unconstrained local maximum with respect to x at x^*. In this chapter, following [L4] and [P7], a particular augmented Lagrangian $L_a(x, \alpha, \beta, w)$ is formed by adding penalty-like terms to L, with w being a set of weighting factors. L_a is formed for the purpose of obtaining direct relationships between constrained local maxima of f(x) and unconstrained local maxima of $L_a(x, \alpha^*, \beta^*, w)$ with respect to x.

After introducing L_a in Section 4.2, definitive properties of L_a are developed by three theorems and a Lemma in Section 4.3. The major features of an augmented-Lagrangian algorithm (a multiplier algorithm) are then presented in Section 4.4, two main phases being: 1) unconstrained maximization of L_a; and 2) multiplier and weight update. The particular multiplier update rule of Section 4.4 is further justified by considerations of local duality (Sections 4.6 and 4.7). The rationale for using a weight update procedure is given in Section 4.8.

4.2 AN AUGMENTED LAGRANGIAN

Consider the augmented Lagrangian L_a:

$$L_a \triangleq L - w_1 P_1 - w_2 P_2 - w_3 P_3 \qquad (4-1)$$

where L is the Lagrangian of (2-58), $w = \{w_1, w_2, w_3\}$ is a set of three penalty weights with each $w_i > 0$, and

4.2 AN AUGMENTED LAGRANGIAN

$$P_1 \triangleq \sum_{i=1}^{m_1} (a_i - p_i)^2 \tag{4-2a}$$

$$P_2 \triangleq \sum_{j \in C_a} (b_j - q_j)^2, \quad C_a = \{j \mid \beta_j > 0\} \tag{4-2b}$$

and

$$P_3 \triangleq \sum_{j \in C_b} (b_j - q_j)^2,$$

$$C_b = \{j \mid \beta_j = 0 \text{ and } q_j \geq b_j\} \tag{4-2c}$$

With slight loss in generality, we could let $w_2 = w_3$ and replace the last two terms in (4-1) by

$$-w_2 \sum_{j \in C_a \cup C_b} (b_j - q_j)^2$$

Note also that $C_a \cap C_b$ equals the null set ϕ, and therefore the $w_3 P_3$ term in (4-1) is equivalent to

$$w_3 \sum_{j \notin C_a} \tfrac{1}{2}(b_j - q_j)[(b_j - q_j) - |b_j - q_j|]$$

In terms of modified Lagrangians, L_a of (4-1) is but one of a wide variety (see Appendix C).

The gradient of L_a with respect to x is of interest:

$$\nabla L_a = \nabla f - \sum_{i=1}^{m_1} \alpha_i^+ \nabla p_i - \sum_{j \in C_a} \beta_j^+ \nabla q_j$$

$$+ \sum_{j \in C_b} 2w_3 (b_j - q_j) \nabla q_j \qquad (4-3)$$

where

$$\alpha_i^+ \triangleq \alpha_i - 2w_1 (a_i - p_i) \qquad (4-4a)$$

and

$$\beta_j^+ \triangleq \beta_j - 2w_2 (b_j - q_j) \qquad (4-4b)$$

Consider the directions of the vectors that sum to form ∇L_a. The first vector ∇f points in the direction of greatest incremental increase in f. A typical vector for an equality constraint is $[-\alpha_i + 2w_1(a_i - p_i)]\nabla p_i$: if $p_i > a_i$, this vector points in the direction of decreasing $|p_i - a_i|$ if $[-\alpha_i + 2w_1(a_i - p_i)] < 0$ or, equivalently, if $p_i - a_i > -\alpha_i/2w_1$; on the other hand, if $p_i < a_i$, the vector points in the direction of decreasing $|p_i - a_i|$ if $[-\alpha_i + 2w_1(a_i - p_i)] > 0$ which is true if $a_i - p_i > \alpha_i/2w_1$. A stronger statement applies for vectors of the form

4.2 AN AUGMENTED LAGRANGIAN

$[-\beta_j + 2w_2(b_j - q_j)]\nabla q_j$: if $q_j > b_j$, this vector points in the direction of decreasing $|q_j - b_j|$ because $\beta_j > 0$; otherwise, if $q_j < b_j$, this vector points in the direction of decreasing $|q_j - b_j|$ if $[-\beta_j + 2w_2(b_j - q_j)] > 0$ which is true if $b_j - q_j > \beta_j/2w_2$. Finally, the vector $2w_3(b_j - q_j)\nabla q_j$, which applies only if $q_j - b_j > 0$, points in the direction of decreasing $|q_j - b_j|$.

4.3 PROPERTIES OF UNCONSTRAINED MAXIMA OF L_a

With the augmented Lagrangian defined in (4-1) there is a direct relationship between the constrained local maxima of the original problem and the unconstrained local maxima of L_a. This relationship is clarified by the three theorems of this section. The first two theorems stem from [P8].

THEOREM 4.1 [Concerning an Unconstrained Local Maximum of $L_a(x, \alpha^*, \beta^*, w)$]
Let (x^*, α^*, β^*) satisfy first-order conditions (2-80). If $L_a(x^*, \alpha^*, \beta^*, w)$ is an unconstrained local maximum of $L_a(x, \alpha^*, \beta^*, w)$ with respect to

x for some finite w > 0, then $f(x^*)$ is a constrained local maximum of $f(x)$.

Proof.

$$L_a(x^* + \Delta x, \alpha^*, \beta^*, w) \leq L_a(x^*, \alpha^*, \beta^*, w) \quad (4-5)$$

for all Δx which satisfy $\|\Delta x\| < \varepsilon$ where ε is some given real value. To simplify notation, we also express (4-5) as $L_a(x^* + \Delta x) \leq L_a(x^*)$ with the understanding that the other arguments of L_a are as in (4-5).

Now, for the constrained problem, only those Δx are allowed for which $x^* + \Delta x$ satisfies the general constraints (2-2) and (2-3). For any such Δx, with $\|\Delta x\| < \varepsilon$, equation (4-1) reduces to

$$L_a(x^* + \Delta x) = f(x^* + \Delta x)$$
$$- \sum_{j \in C_a} s_j(x^* + \Delta x) \cdot$$
$$[\beta_j^* + w_2 s_j(x^* + \Delta x)] \quad (4-6)$$

in which

$$s_j(x^* + \Delta x) \triangleq q_j(x^* + \Delta x) - b_j \quad (4-7)$$

4.3 AN AUGMENTED LAGRANGIAN 133

so that $s_j(x^* + \Delta x) \leq 0$ for allowed Δx's, and $s_j(x^*) = 0$.

Consider a typical term

$$s_j(x^* + \Delta x)[\beta_j^* + w_2 s_j(x^* + \Delta x)] \qquad (4-8)$$

from the summation in (4-6). With $s_j(x^* + \Delta x) \leq 0$, $\beta_j^* > 0$ for $j \in C_a$, and $0 < w_2 < \infty$, there exists an $\varepsilon > 0$ such that, when $\|\Delta x\| < \varepsilon$, $\beta_j^* > -w_2 s_j(x^* + \Delta x)$; under this condition, (4-8) is nonpositive. Thus, in (4-6), nonpositive terms are subtracted from $f(x^* + \Delta x)$ to form $L_a(x^* + \Delta x)$, and therefore,

$$L_a(x^* + \Delta x) \geq f(x^* + \Delta x) \qquad (4-9)$$

for allowed Δx's. By assumption, $L_a(x^*)$ is an unconstrained local maximum of L_a, and thus

$$f(x^*) = L_a(x^*) \geq L_a(x^* + \Delta x) \qquad (4-10)$$

and from (4-9) and (4-10), therefore,

$$f(x^*) \geq f(x^* + \Delta x) \qquad (4-11)$$

for allowed Δx's. Thus, $f(x^*)$ is a constrained local maximum of f.

In the above proof if the inequalities in (4-10), and therefore also in (4-11), can be replaced by strict inequalities, then in Theorem 4.1 it can be stated that $f(x^*)$ is a strict constrained local maximum if $L_a(x, \alpha^*, \beta^*, w)$ assumes a strict local maximum with respect to x at x^*.

EXAMPLE 4.1 Consider again the area-volume problem of Example 2.2. The appropriate augmented Lagrangian is

$$L_a = L - w_1 \left(\frac{x_1^2}{2} + x_1 x_2 - a_1 \right)^2 \qquad (4\text{-}12)$$

The corresponding gradient is

$$\nabla L_a = \nabla L - \begin{bmatrix} x_1 + x_2 \\ \\ x_1 \end{bmatrix} 2 w_1 \left(\frac{x_1^2}{2} + x_1 x_2 - a_1 \right) \qquad (4\text{-}13)$$

and

4.3 AN AUGMENTED LAGRANGIAN

$$\nabla^2 L_a = \nabla^2 L - 2w_1 \begin{bmatrix} (x_1+x_2)^2 & x_1(x_1+x_2) \\ x_1(x_1+x_2) & x_1^2 \end{bmatrix}$$

$$- \begin{bmatrix} 1 & 1 \\ 1 & 0 \end{bmatrix} 2w_1 \left(\frac{x_1^2}{2} + x_1 x_2 - a_1 \right) \quad (4\text{-}14)$$

At $x = x^* = [\alpha_1^* \; \alpha_1^*]'$ as determined in Example 2.2,

$$\nabla L_a^* = \nabla L^* = 0 \quad (4\text{-}15)$$

and

$$\nabla^2 L_a^* = \begin{bmatrix} \alpha_1^* & \alpha_1^* \\ \alpha_1^* & 0 \end{bmatrix} - 2w_1 \begin{bmatrix} 4\alpha_1^{*2} & 2\alpha_1^{*2} \\ 2\alpha_1^{*2} & \alpha_1^{*2} \end{bmatrix}$$

$$= \begin{bmatrix} -8w_1\alpha_1^{*2} + \alpha_1^* & -4w_1\alpha_1^{*2} + \alpha_1^* \\ -4w_1\alpha_1^{*2} + \alpha_1^* & -2w_1\alpha_1^{*2} \end{bmatrix} \quad (4\text{-}16)$$

By Sylvester's test (Appendix D), $-\nabla^2 L_a^*$ is positive definite if

$$8w_1 \alpha_1^{*2} - \alpha_1^* > 0 \quad \text{and}$$

$$2w_1\alpha_1^{*2}(8w_1\alpha_1^{*2} - \alpha_1^*) - (4w_1\alpha_1^{*2} - \alpha_1^*)^2 > 0 \qquad (4\text{-}17)$$

Both inequalities are satisfied if

$$w_1 > \frac{1}{6\alpha_1^*} = \frac{1}{6}\left(\frac{3}{2a_1}\right)^{\frac{1}{2}} \qquad (4\text{-}18)$$

Thus, $-\nabla^2 L_a^*$ is positive definite for sufficiently large but finite w. From Theorem 4.1, therefore, x^* is also a point of constrained local maximum of $f(x)$.

From the above example, it would appear that the converse of Theorem 4.1 also holds. Unfortunately, this is not always true, as is shown by the following example which is equivalent to an example supplied by Rockafellar [R3].

EXAMPLE 4.2 Let

$$f = x_2 - x_1^4 - x_1 x_2 \qquad (4\text{-}19)$$

which is to be maximized subject to the constraint that

$$p_1(x) = x_2 = 0 \qquad (4\text{-}20)$$

4.3 AN AUGMENTED LAGRANGIAN

The Lagrangian is given by

$$L = (1 - \alpha_1)x_2 - x_1^4 - x_1 x_2 \qquad (4-21)$$

and

$$\nabla L = [-(4x_1^3 + x_2) \quad (1 - \alpha_1 - x_1)]' \qquad (4-22)$$

From (4-22), first-order necessary conditions are satisfied by $x_1^* = x_2^* = 0$ and $\alpha_1^* = 1$. That $x_1^* = x_2^* = 0$ is also the constrained maximum point is clearly evident from (4-19) and (4-20).

However, the augmented Lagrangian is

$$L_a(x, \alpha_1^*) = -x_1^4 - x_1 x_2 - w_1 x_2^2 \qquad (4-23)$$

which exhibits a saddle point at $(0, 0)$ and maxima at $\pm((8w_1)^{-\frac{1}{2}}, -(32w_1^3)^{-\frac{1}{2}})$ (see Problem 4.6).

In lieu of the converse of Theorem 4.1, we have Theorem 4.2.

THEOREM 4.2 [Relating Constrained Local Maxima of $f(x)$ to Unconstrained Local Maxima of $L_a(x, \alpha^*, \beta^*, w)$]

Let (x^*, α^*, β^*) satisfy first-order conditions (2-80). If $f(x^*)$ is a constrained local maximum of $f(x)$, then either:

i) $L_a(x^*, \alpha^*, \beta^*, w)$ is an unconstrained local maximum of $L_a(x, \alpha^*, \beta^*, w)$ with respect to x for some finite $w > 0$; or

ii) in the limit as w_i's $\to \infty$, $L_a(x^*, \alpha^*, \beta^*, w)$ approaches an unconstrained local maximum of $L_a(x, \alpha^*, \beta^*, w)$.

<u>Proof</u>. First, consider any $x^* + \Delta x$ which does not satisfy all of the constraints: any sufficiently large w_i's can be used in (4-1) to force $L_a(x^* + \Delta x)$ to be less than $L_a(x^*)$, with the possibility that one or more w_i may have to approach infinity. For the same reason, if $x^* + \Delta x$ satisfies all of the constraints, but one or more $q_j(x^* + \Delta x) < b_j$ with $j \in C_a$, $L_a(x^* + \Delta x)$ is less than $L_a(x^*)$ for sufficiently large w_2.

Now, let a set of Δx's be defined by the following:

$$p_i(x^* + \Delta x) = a_i, \quad i = 1, 2, \ldots, m_1 \quad (4\text{-}24a)$$

$$q_j(x^* + \Delta x) = b_j, \quad \text{for } j \in C_a \quad (4\text{-}24b)$$

$$q_j(x^* + \Delta x) \leq b_j, \quad \text{for } j \notin C_a \quad (4\text{-}24c)$$

and

4.3 AN AUGMENTED LAGRANGIAN

$$\|\Delta x\| < \varepsilon \qquad (4\text{-}24d)$$

such that $f(x^* + \Delta x) \leq f(x^*)$. For any Δx that does not satisfy (4-24), the preceding paragraph gives $\lim_{w \to \infty} L_a(x^* + \Delta x) < L_a(x^*)$. But for any Δx which is in the set defined by (4-24),

$$L_a(x^* + \Delta x) = f(x^* + \Delta x) \leq f(x^*) = L_a(x^*) \qquad (4\text{-}25)$$

Thus, either $L_a(x^*)$ is a local unconstrained maximum of L_a for sufficiently large w, or it approaches such a maximum as $w \to \infty$.

●

If the inequality in (4-25) can be replaced by a strict inequality (that is, if $f(x^*)$ is a strict constrained local maximum of f), then either $L_a(x^*)$ is a strict local maximum point for L_a for sufficiently large w, or it approaches such a maximum as $w \to \infty$.

Statement i of Theorem 4.2 holds in Example 4.1. In Example 4.2, however, statement ii applies: points of local maxima are located at $x = \pm[(8w_1)^{-\frac{1}{2}}, -(32w_1^3)^{-\frac{1}{2}}]$ and approach the constrained optimum point $(0, 0)$ as $w_1 \to \infty$.

Statement i of Theorem 4.2 is clearly preferable to statement ii. It is fortunate, therefore, that statement i applies to a broad class of problems, as is shown by the following theorem [L4, P8].

THEOREM 4.3 [Equivalence Between NLP Solutions and Unconstrained Maxima of L_a; Finite Weights Guaranteed]
Given (x^*, α^*, β^*) that satisfy (2-80) and given sufficiently large but finite w, $L_a(x, \alpha^*, \beta^*, w)$ satisfies second-order sufficient conditions for an unconstrained local maximum at x^*, if and only if the NLP satisfies second-order sufficient conditions for a constrained local maximum at x^*, α^*, β^*.

Before proving Theorem 4.3, we state and prove Lemma 4.1 which is related to a theorem (Theorem 3 of [D5]) on quadratic forms. Notation used in the Lemma is as follows: let B be an n×m matrix of reals and let r_j be an n×1 column vector of reals for each j contained in a given integer set S. Further, we introduce the sets V and $\hat{V}$:

4.3 AN AUGMENTED LAGRANGIAN

$$V \triangleq \{\Delta x \mid B'\Delta x = 0, r_j'\Delta x \leq 0 \text{ for } j \in S,$$
$$\text{and } \Delta x \neq 0\} \quad (4\text{-}26a)$$

$$\hat{V} \triangleq \{\Delta x \mid \Delta x \notin V \text{ and } \Delta x \neq 0\} \quad (4\text{-}26b)$$

Also,

$$\delta_j \triangleq \begin{cases} 1, & \text{if } r_j'\Delta x > 0 \\ & \\ 0, & \text{if } r_j'\Delta x \leq 0 \end{cases}, \quad j \in S \quad (4\text{-}27)$$

LEMMA 4.1 [Quadratic Forms and Linear Equalities and Inequalities]

Let B, V, $\hat{V}$, and δ_j's be defined as above and let A be an n×n symmetric matrix of reals. Then for $\Delta x \in V$, $\Delta x'A\Delta x < 0$ if and only if there exists a number λ such that, for all $\Delta x \neq 0$,

$$F \triangleq \Delta x'A\Delta x - \lambda \Delta x'(BB' + \sum_{j \in S} r_j r_j' \delta_j)\Delta x < 0 \quad (4\text{-}28)$$

Proof. First suppose $F < 0$ for all $\Delta x \neq 0$. For $\Delta x \in V$, $F = \Delta x'A\Delta x < 0$ which proves the sufficient part of the Lemma.

Next, suppose that $\Delta x'A\Delta x < 0$ for $\Delta x \in V$. For these Δx, it is obvious that $F < 0$ for any finite λ. For $\Delta x \in \hat{V}$, we use the following argument: let

$$\theta(\Delta x) \triangleq \frac{\Delta x' A \Delta x}{\Delta x'(BB' + \sum_{j \in S} r_j r_j' \delta_j)\Delta x} \qquad (4\text{-}29)$$

Because of the way in which Δx enters into the right-hand side of (4-29), we can restrict our attention to Δx's contained in normalized sets V_n and $\hat{V}_n$, where

$$V_n \triangleq \{\Delta x \mid \Delta x \in V \text{ and } \|\Delta x\| = 1\} \qquad (4\text{-}30a)$$

and

$$\hat{V}_n \triangleq \{\Delta x \mid \Delta x \in \hat{V} \text{ and } \|\Delta x\| = 1\} \qquad (4\text{-}30b)$$

The denominator of (4-29) is of particular interest; that is, $\Delta x'(BB' + \sum_{j \in S} r_j r_j' \delta_j)\Delta x$ exhibits the following properties: 1) it is zero for $\Delta x \in V_n$; 2) it is a continuous function of Δx; and 3) it is greater than zero for $\Delta x \in \hat{V}_n$. Thus, if we allow a $\Delta x \in \hat{V}_n$ to approach the boundary between V_n and $\hat{V}_n$, the denominator of (4-29) approaches 0+. The numerator $\Delta x' A \Delta x$ of (4-29), on the other hand, is assumed to be strictly less than zero on the boundary. We see, therefore, that $\theta(\Delta x) \to -\infty$ as $\Delta x \in \hat{V}_n$ approaches the boundary

4.3 AN AUGMENTED LAGRANGIAN 143

of $\hat{V}_n$. Inside $\hat{V}_n$, however, $\theta(\Delta x)$ is continuous. We conclude that $\theta(\Delta x)$ assumes a finite maximum value λ^* at some point in $\hat{V}_n$. Thus, for any $\lambda > \lambda^*$,

$$\lambda > \frac{\Delta x' A \Delta x}{\Delta x' (BB' + \sum_{j \in S} r_j r_j' \delta_j) \Delta x}, \quad \Delta x \in \hat{V}_n \quad (4-31)$$

By rearranging (4-31), we obtain the desired result that

$$\Delta x' A \Delta x - \lambda \Delta x' (BB' + \sum_{j \in S} r_j r_j' \delta_j) \Delta x < 0,$$

$$\Delta x \in \hat{V}_n \quad (4-32)$$

which, when combined with those $\Delta x \in V$, yields (4-28) for all $\Delta x \neq 0$, thus proving the necessary part of the Lemma. ●

We may now utilize the results of Lemma 4.1 in proving Theorem 4.3.

Proof. (Theorem 4.3) Under the assumed conditions of the theorem, it is clear that $\nabla L_a^* = \nabla L^* = 0$. It must be shown that, for all $\Delta x \neq 0$ and finite w_1, w_2, and w_3, $\Delta x' \nabla^2 L_a^* \Delta x < 0$ if and only if $\Delta x' \nabla^2 L^* \Delta x < 0$ for $\Delta x \neq 0$ and Δx such that

$\Delta x'\nabla p_i^* = 0$, $i = 1, 2, \ldots, m_1$, and $\Delta x'\nabla q_j^* = 0$, $j \in C_a$, and $\Delta x'\nabla q_j^* \leq 0$, $j \in S_a - C_a$ at x^*.

The proof follows directly from Lemma 4.1 when the following identities are made. Without loss of generality, let $\lambda = w_1 = w_2 = w_3$ and

$$F \equiv \tfrac{1}{2}\Delta x'\nabla^2 L_a^* \Delta x \tag{4-33a}$$

$$A \equiv \tfrac{1}{2}\nabla^2 L^* \tag{4-33b}$$

$$\text{Columns of } B \begin{cases} \nabla p_i^*, \ i = 1, 2, \ldots, m_1 \\ \\ \nabla q_j^*, \ j \in C_a \end{cases} \tag{4-33c}$$

and

$$r_j \equiv \nabla q_j^*, \ j \in S = S_a - C_a \tag{4-33d}$$

●

4.4 AN OVERVIEW OF A MULTIPLIER ALGORITHM

Our concern to this point in Chapter 4 has been $L_a(x, \alpha^*, \beta^*, w)$. In using L_a to solve an NLP, however, optimal multipliers α^* and β^* are seldom available in advance, so that $L_a(x, \alpha, \beta, w)$ must be operated upon in some systematic way

4.4 AN AUGMENTED LAGRANGIAN 145

to find consistent values of x^* and (α^*, β^*) which satisfy optimality conditions. Although there are many possible methods of doing this, we focus our attention here on the main features of a straightforward approach which has proved to be highly effective; this approach and variations of it are encoded in the computer subroutines of Appendix B.

Because of the penalty-like terms included in L_a, unconstrained maxima of L_a generally exist for finite w, so we start with fixed w > 0 and with fixed values of α and β (without advance information, $\alpha = \beta = 0$ may be used at the start) and generate an unconstrained maximum of L_a. Any number of good unconstrained maximization techniques can be used for this purpose; several of these are described in Appendix E.

In finding the first unconstrained maximum point x^1, we obtain

$$\nabla L_a(x^1) \cong 0 \qquad (4-34)$$

where x^1, in general, does not satisfy all constraints. Also, the optimality conditions require $\nabla L = 0$, rather than $\nabla L_a = 0$. In order to satisfy

$\nabla L(x^1) \cong 0$, the next phase of the algorithm consists of reassigning α and β so that the new $\nabla L(x^1)$ equals the old $\nabla L_a(x^1)$; the way this is done is clarified in the next few pages. Also, the rationale for updating the multipliers in this way is given a broader base in Sections 4.6 and 4.7 by the concept of local duality. After the multipliers are updated, the weights w_1, w_2, and w_3 may also be increased, and when second-order sufficient conditions are not satisfied, they may have to be increased for the algorithm to generate a valid solution.

After the multipliers and weights are updated, the process is repeated: the algorithm consists of alternating unconstrained maximization phases and update phases. During the mth maximization phase, the multipliers and penalty weights are held fixed, and a quadratically convergent search is employed to find x^m that maximizes $L_a(x)$ such that x^m satisfies

$$\nabla L_a(x^m) = 0 \qquad (4-35)$$

Because of computational roundoff errors and

4.4 AN AUGMENTED LAGRANGIAN

because nonlinear functions are involved, x^m generally satisfies (4-35) only in an approximate sense: the termination of the unconstrained maximization phase must be based on a set of pragmatic conditions. One condition of the set is as follows: let ε be a small number and replace (4-35) by

$$\nabla L_a(x^m)'\nabla L_a(x^m) < \varepsilon \qquad (4-36)$$

At the completion of the mth maximization phase, the multipliers are updated in an attempt to satisfy $\nabla L = 0$. Thus, the multipliers are updated to obtain

$$\underset{\text{(new)}}{\nabla L(x^m)} = \underset{\text{(old)}}{\nabla L_a(x^m)} \simeq 0 \qquad (4-37)$$

The multiplier updating rules that satisfy (4-37) follow by comparing ∇L of (2-59) with ∇L_a of (4-3):

For equality constraints:

$$\alpha_i^+ = \alpha_i - 2w_1[a_i - p_i(x^m)],$$

$$i = 1, 2, \ldots, m_1 \qquad (4-38)$$

where α_i^+ denotes the new value of α_i, and the old value of α_i is used in the right-hand side of (4-38).

For inequality constraints:

If $j \in C_a$,

$$\beta_j^+ = \begin{cases} 0, & \text{if } \beta_j - 2w_2[b_j - q_j(x^m)] \leq 0 \\ \beta_j - 2w_2[b_j - q_j(x^m)], & \text{otherwise} \end{cases} \quad (4-39)$$

If $j \in C_b$,

$$\beta_j^+ = \begin{cases} 0, & \text{if } b_j - q_j(x^m) \geq 0 \\ -2w_3[b_j - q_j(x^m)], & \text{otherwise} \end{cases} \quad (4-40)$$

Here also, β_j^+ denotes the new value of β_j.

The w_i values used in (4-38), (4-39), and (4-40) are those values which were used during the preceding maximization phase. After the new β_j's are generated, the indices belonging to the sets C_a and C_b are updated, and are held fixed for the next maximization phase.

Equations (4-39) and (4-40) prescribe a <u>constraint basis</u> for the inequality constraints

4.4 AN AUGMENTED LAGRANGIAN 149

during a maximization phase. Throughout such a maximization phase, those inequalities having positive multipliers are in a <u>locked-on</u> status and are treated the same as equality constraints. Inequality constraints with associated zero-valued multipliers are held from excessive constraint violation, as in penalty function methods, by the exterior penalty terms associated with w_3. Upon termination of a given maximization phase, the equality constraint multipliers are updated by (4-38), and the inequality constraint multipliers are updated by (4-39) and (4-40) in order to prescribe a new inequality constraint basis for the next maximization phase.

If a β_j multiplier from the previous basis tends negative, as in (4-39), the implication is that f will tend to increase if x is free to move away from the corresponding constraint boundary with $b_j - q_j(x)$ increasing, thus relinquishing its locked-on status. The constraint then becomes a <u>free constraint</u>. A free constraint is added to the new constraint basis during the update phase if the constraint is violated at the end of a maximization phase. Thus, with the violated constraint $b_j - q_j(x)$ having a value less than zero, the corresponding multiplier is assigned a positive value, as in (4-40), and the constraint maintains a locked-on status throughout the next maximization phase.

The locked-on feature is especially important when x^m is in the vicinity of a constrained maximum at x^*. If the problem functions are of class C^2 and if <u>strict complimentary slackness</u> holds near x^* (i.e., if $S_a = C_a$), all active inequality constraints (characterized by $q_j(x^*) = b_j$) have associated positive multipliers ($\beta_j > 0$), and the matrix of second partial derivatives $\nabla^2 L_a^*$ is continuous. Continuity of second derivatives is a property that facilitates the unconstrained maximization of a function by the use of quadratic convergent techniques.

For most problems, the w_i's could be assigned initially and held fixed throughout the algorithm. But because the terms added to the Lagrangian to form L_a are penalty terms, it is more logical to parallel to some degree the procedure used with

4.4 AN AUGMENTED LAGRANGIAN

penalty function methods (more is said on this in Section 4.8). Thus, the w_i's are assigned some relatively small positive values and are updated, after the multipliers are updated, by a factor $\gamma \geq 1$. This procedure continues until given upper bounds w_{imax}'s are reached.† The penalty weight update rule then is

$$w_i^+ = \begin{cases} w_{imax}, & \text{if } \gamma w_i \geq w_{imax} \\ & \qquad\qquad i = 1, 2, 3 \quad (4\text{-}41) \\ \gamma w_i, & \text{otherwise} \end{cases}$$

where w_i^+ denotes the new value of w_i.

It is important to note that the initial w_i's must be large enough to prevent a constraint breakthrough from occurring. A <u>constraint breakthrough</u> occurs when an unconstrained search for a maximum of L_a generates a sequence of x's with the property that f(x) increases at a rate that is greater than the rate that the $-w_i P_i$ penalty terms can decrease, thereby leading to an unbounded and nonfeasible solution. For example, the presence of

---

† The initial w_i's and the upper bounds are assigned by the user, as described in Appendix A.

exponentials such as $\exp(x_i)$ in $f(x)$ tends to foster occurrences of constraint breakthroughs.

At x^m, after the multipliers and weights have been updated, the search for a maximum of L_a is conducted initially in the $\nabla L_a(x^m)$ direction[†], where

$$\nabla L_a(x^m)\Big|_{(\alpha,\beta,w) = (\alpha^+,\beta^+,w^+)} = [2w_1 \sum_{i=1}^{m_1} (a_i - p_i)\nabla p_i + 2w_2 \sum_{j \in C_a} (b_j - q_j)\nabla q_j]_{x=x^m} \quad (4\text{-}42)$$

The search direction of (4-42) is one that tends to move x towards the boundaries of the active constraints, and in general, the corresponding constraints will be more closely satisfied at the end of the next maximization phase.

4.5 FINITE CONVERGENCE FOR A GENERAL LINEAR CASE

The preceding algorithm generates the constrained maximum of f after a <u>finite number</u> of

---

[†] This is true only for certain of the options given in the program of Appendices A and B.

4.5 AN AUGMENTED LAGRANGIAN

unconstrained quadratic-convergent maximizations under the following conditions:

1) the p_i's, q_j's, and f are linear in x,

2) a unique and bounded x^* exists, and

3) $q_j(x^*) < b_j$ if $\beta_j^* = 0$.

Justification of the above is as follows. At some finite point in the sequence of alternating maximization and update phases, the constraint basis C_a contains only those constraints that bound x^* and the set C_b is empty. A quadratic-convergent maximization is then performed where the last summation in (4-1) remains at zero. The resulting x^m satisfies $\nabla L_a(x^m) = 0$, apart from roundoff errors, because L_a is quadratic in the domain leading from x^{m-1} to x^m. The multipliers obtained at this stage are optimum because the gradients of the linear problem functions are constant vectors. Thus, the new $\nabla L = 0$ and the new ∇L_a is that given in (4-42), with the terms associated with w_1 and w_2 quadratic in x^m. One final quadratic-convergent maximization phase yields the true maximum within numerical accuracy limitations.

4.6 LOCAL DUALITY

Consider the standard problem:

$$\text{maximize } f(x) \qquad (4\text{-}43a)$$

subject to

$$p(x) = a \qquad (4\text{-}43b)$$

and

$$q(x) \leq b \qquad (4\text{-}43c)$$

with the usual assumptions in regard to f, p, and q. We restrict our attention to a local solution x^*, which is both a regular point of the constraints and a point at which C_b of (4-2c) is empty. Multipliers exist, therefore, with the property that

$$\nabla L_a^* = \nabla L^* = \nabla f^* - (\alpha^*)'\nabla p^* - (\beta^*)'\nabla q^* = 0 \qquad (4\text{-}44)$$

where $\beta_j^* q_j^* = 0$ and $\beta_j^* \geq 0$ for $j = 1, 2, \ldots, m_2$. We also assume that second-order sufficient

4.6 AN AUGMENTED LAGRANGIAN

conditions are satisfied at x^*, and are therefore assured by Theorem 4.3 that $L_a(x, \alpha^*, \beta^*, w)$ exhibits an unconstrained local maximum at x^* for sufficiently large but finite penalty weights w_1, w_2, and w_3.

Without loss of generality, let $w_1 = w_2 = w_3$ in this development of local duality, and let the q_j's and associated β_j's be ordered such that

$$q_j(x^*) = b_j$$
$$\qquad\qquad j = 1, 2, \ldots, r \leq m_2 \qquad (4\text{-}45)$$
$$\beta_j^* > 0$$

and

$$q_j(x^*) < b_j$$
$$\qquad\qquad j = r+1, r+2, \ldots, m_2 \qquad (4\text{-}46)$$
$$\beta_j^* = 0$$

The arrangements in (4-45) and (4-46) are consistent with the assumption that C_b is empty.

The following notation is introduced for ease in presentation. Let

$$\eta \triangleq [\alpha_1 \; \alpha_2 \; \cdots \; \alpha_{m_1} \; \beta_1 \; \beta_2 \; \cdots \; \beta_r]' \qquad (4\text{-}47)$$

and

$$s(x) \triangleq [(p_1 - a_1) \cdots (p_{m_1} - a_{m_1})$$
$$(q_1 - b_1) \cdots (q_r - b_r)]'$$
$$= [s_1 \; s_2 \; \cdots \; s_{m_1+r}]' \qquad (4\text{-}48)$$

The <u>reduced augmented Lagrangian</u> is then defined, as follows:

$$L_r(x, \eta) \triangleq f - \eta's - w_1 s's \qquad (4\text{-}49)$$

Note that L_r differs from L_a: inequality constraints that are inactive at x^* are not included in L_r, and the multipliers associated with these inactive constraints are also not germane to L_r. However, both L_a and L_r have the property that those active inequality constraints for which corresponding β_j's are strictly greater than zero are in a locked-on status; they enter into L_r in the same way as do equality constraints. Also, we are given that x^* is a particular type of constrained local maximum, and from Theorem 4.3, we know that $L_a(x, \alpha^*, \beta^*, w)$ exhibits an unconstrained local maximum at $x = x^*$; it readily follows that an

4.6 AN AUGMENTED LAGRANGIAN

unconstrained local maximum of $L_r(x,\eta^*)$ exists at $x = x^*$.

Consider a value of η close to η^*. The Lagrangian $L_r(x,\eta)$ will have a local maximum near x^* because

$$\nabla L_r(x,\eta) = 0 \qquad (4\text{-}50)$$

has a solution near x^* when η is near η^* and because second-order sufficient conditions for a maximum of $L_r(x,\eta^*)$ are known to hold at x^*.

Near η^*, the <u>local dual function</u> $D(\eta)$ is defined by

$$D(\eta) \triangleq \text{maximum } L_r(x,\eta) \qquad (4\text{-}51)$$

in which the maximization is taken with respect to x in the neighborhood of x^*. In (4-51), let $x(\eta)$ denote the x that yields the maximum. Then

$$D(\eta) = L_r(x(\eta),\eta) \qquad (4\text{-}52)$$

Geoffrion [G2] and Vajda [V1] give historical accounts of such dual functions.

Of interest are the properties of $D(\eta)$ in (4-52) when η is in the neighborhood of η^*. An appropriate series expansion for $D(\eta)$ is

$$D(\eta) = D(\eta^*) + \Delta\eta' \nabla D(\eta^*)$$
$$+ \tfrac{1}{2}\Delta\eta' \nabla^2 D(\eta^*)\Delta\eta + \cdots \qquad (4-53)$$

where $\Delta\eta \triangleq \eta-\eta^*$. To fully appreciate (4-53), meaningful expressions for ∇D and $\nabla^2 D$ are required.

To obtain $\nabla D(\eta)$, (4-52) is used to form

$$\nabla D(\eta) = \frac{\partial L_r(x(\eta),\eta)}{\partial \eta}$$
$$+ \left[\frac{dx(\eta)}{d\eta}\right]' \frac{\partial L_r(x(\eta),\eta)}{\partial x} \qquad (4-54)$$

but because $[\partial L_r(x(\eta),\eta)/\partial x] = 0$ by the nature of D and the definition of $x(\eta)$, it follows that

$$\nabla D(\eta) = \frac{\partial L_r}{\partial \eta} = -s(x(\eta)) \qquad (4-55)$$

Note that (4-55) gives the gradient of the local dual function directly in terms of the constraint functions which are readily evaluated.

The Hessian $\nabla^2 D$ of the dual function is

$$\nabla^2 D = \left[\frac{d^2 D}{d\eta_i d\eta_j}\right] = \frac{d}{d\eta} \nabla D \qquad (4-56)$$

4.6 AN AUGMENTED LAGRANGIAN

From (4-55) and (4-56),

$$\nabla^2 D = -\frac{d}{d\eta} s(x(\eta)) = -\frac{ds}{dx}\frac{dx(\eta)}{d\eta} = -\nabla s \nabla x \tag{4-57}$$

with the understanding that $\nabla s = [ds(x)/dx]_{x=x(\eta)}$ and $\nabla x = dx(\eta)/d\eta$. Expression (4-57) for $\nabla^2 D$ can be made more meaningful if we can find a relevant expression for $\nabla x(\eta)$. To this end, recall that the gradient $\nabla L_r(x(\eta),\eta)$ of L_r with respect to x is identically zero, in which case, the total derivative of $\nabla L_r(x(\eta),\eta)$ with respect to η is also identically zero; that is

$$\frac{d}{d\eta}[\nabla L_r(x(\eta),\eta)] = [\frac{\partial}{\partial x} \nabla L_r(x(\eta),\eta)]\nabla x(\eta)$$

$$+ \frac{\partial}{\partial \eta} \nabla L_r(x(\eta),\eta)$$

$$= \nabla^2 L_r(x(\eta),\eta)\nabla x(\eta)$$

$$- \nabla s(x(\eta))´ = 0 \tag{4-58}$$

in which $\nabla^2 L_r$ is given by

$$\nabla^2 L_r = \nabla^2 f - \sum_{i=1}^{m_1+r} [(n_i + 2w_1 s_i)\nabla^2 s_i$$

$$+ 2w_1 \nabla s_i \nabla s_i´] \tag{4-59}$$

Now, by the nature of x^*, we know that $L_r(x,\eta^*)$ assumes an unconstrained local maximum at x^* and, therefore, that $-\Delta x' \nabla^2 L_r(x(\eta),\eta) \Delta x > 0$ for $\Delta x = x(\eta) - x^* \neq 0$ when η is near η^*. It must be true then that $[\nabla^2 L_r(x(\eta),\eta)]^{-1}$ exists, so (4-58) can be rearranged to obtain

$$\nabla x(\eta) = [\nabla^2 L_r(x(\eta),\eta)]^{-1} \nabla s' \qquad (4-60)$$

The right-hand member of (4-60) is substituted into (4-57) with the desired result that

$$\nabla^2 D = -\nabla s [\nabla^2 L_r(x(\eta),\eta)]^{-1} \nabla s' \qquad (4-61)$$

Before giving the local duality theorem, we state formally the Primal Problem and the Dual Problem, in terms of the notation introduced in this section.

Primal Problem: Maximize $f(x)$, subject to $p(x) = a$ and $q(x) \leq b$, where f, p, and q are assumed to be of class C^2.

Dual Problem: Minimize $D(\eta)$, where $D(\eta)$ is an unconstrained maximum of $L_r(x,\eta)$ with respect to x.

4.6 AN AUGMENTED LAGRANGIAN

THEOREM 4.4 [Local Duality Theorem]

Given the following: 1) the Primal Problem has a local maximum point x^* with value $f(x^*)$; 2) x^* is a regular point of the active constraints and therefore optimal multiplier vectors α^* and β^* exist; 3) nonzero multipliers are associated with the inequality constraints which are active at x^*; and 4) second-order sufficient conditions are satisfied at x^*. Under these conditions, the local dual function $D(\eta)$ exhibits an unconstrained local minimum at $\eta = \eta^*$, where $D(\eta)$ is defined in terms of the reduced augmented Lagrangian in (4-51), and where η^* consists of α^* and non-zero values of β^*. Also $D(\eta^*) = f(x^*)$.

<u>Proof</u>. The fact that x^* corresponds to η^* follows from the definition of D in (4-51). At $\eta = \eta^*$, equation (4-55) gives

$$\nabla D(\eta^*) = -s(x(\eta^*)) = -s(x^*) = 0 \qquad (4\text{-}62)$$

This result and equation (4-61) are used in (4-53) to obtain

$$D(\eta) - D(\eta^*) = -\tfrac{1}{2}\Delta\eta'\nabla s(x^*)[\nabla^2 L_r(x^*,\eta^*)]^{-1} \cdot$$
$$\nabla s(x^*)'\Delta\eta + \cdots \qquad (4\text{-}63)$$

162   MATH PROGRAMMING VIA AUGMENTED LAGRANGIANS

In (4-63), the rows of $\nabla s(x^*)$ are the gradient vectors of the s_i's and are linearly independent by the regularity assumption; and therefore, $\|\Delta\eta'\nabla s(x^*)\| = 0$ if and only if $\|\Delta\eta\| = 0$. Also we know that $-\Delta x' \nabla^2 L_r(x^*,\eta^*)\Delta x > 0$ for $\Delta x \neq 0$, from which it follows that $-\Delta x'[\nabla^2 L_r(x^*,\eta^*)]^{-1}\Delta x > 0$ for $\Delta x \neq 0$. Thus, (4-63) gives that $D(\eta) - D(\eta^*) > 0$ for sufficiently small $\|\Delta\eta\| \neq 0$, and η^* is therefore a local minimum point for $D(\eta)$.

●

The reduced augmented Lagrangian $L_r(x,\eta)$ of the local duality theorem exhibits a <u>saddle point</u> at (x^*,η^*) because it satisfies (see Problem 4.15)

$$L_r(x,\eta^*) \leq L_r(x^*,\eta^*) \leq L_r(x^*,\eta) \qquad (4\text{-}64)$$

provided that the pair (x,η) is in a sufficiently small neighborhood of (x^*,η^*). In other words, there exists an $\varepsilon > 0$ such that (4-64) is satisfied whenever $\|x - x^*\| \leq \varepsilon$ and $\|\eta - \eta^*\| \leq \varepsilon$.

EXAMPLE 4.3   Reconsider the area-volume problem of Examples 2.2 and 4.1. Here L_r equals L_a because only one equality constraint applies. From

4.6 AN AUGMENTED LAGRANGIAN

(2-32) and (4-13), it can be shown that ∇L_a equals zero when

$$x_1 = x_2 = z(\alpha_1) \qquad (4\text{-}65)$$

where $z = z(\alpha_1)$ satisfies

$$z - \alpha_1 - w_1(3z^2 - 2a_1) = 0 \qquad (4\text{-}66)$$

Then, from (4-12) and (4-65),

$$\begin{aligned} D(\alpha_1) &= L_a(x(\alpha_1),\alpha_1) \\ &= z^3 + \alpha_1(a_1 - 1.5z^2) \\ &\quad - w_1(1.5z^2 - a_1)^2 \end{aligned} \qquad (4\text{-}67)$$

from which

$$\frac{dD(\alpha_1)}{d\alpha_1} = [3z^2 - \alpha_1 3z - 6w_1 z(1.5z^2 - a_1)]\frac{dz}{d\alpha_1}$$

$$+ a_1 - 1.5z^2 \qquad (4\text{-}68)$$

From (4-66), the term within brackets in (4-68) is identically zero; and also from (4-66), the remaining term, $a_1 - 1.5z^2$, equals $(\alpha_1 - z)/2w_1$. Thus,

$$\frac{dD(\alpha_1)}{d\alpha_1} = \frac{\alpha_1 - z(\alpha_1)}{2w_1} \qquad (4\text{-}69)$$

and

$$\frac{d^2D}{d\alpha_1^2} = \frac{(1 - dz/d\alpha_1)}{2w_1} \qquad (4\text{-}70)$$

Now, from local duality, $dD(\alpha^*)/d\alpha = 0$, which is satisfied by (4-69) when $z(\alpha_1^*) = \alpha_1^*$, and therefore from (4-66) when $\alpha_1^* = (2a_1/3)^{\frac{1}{2}}$, which checks with the result (2-35) obtained previously. The minimizing nature of α_1^* can be established by examining $d^2D^*/d\alpha_1^2$. From (4-66),

$$\frac{dz}{d\alpha_1} = \frac{1}{1 - 6w_1 z(\alpha_1)} \qquad (4\text{-}71)$$

which is used in (4-70) to obtain

$$\frac{d^2D}{d\alpha_1^2} = \frac{3z(\alpha_1)}{6w_1 z(\alpha_1) - 1} \qquad (4\text{-}72)$$

and at $z(\alpha_1) = \alpha_1^*$,

$$\frac{d^2D^*}{d\alpha_1^2} = \frac{3\alpha_1^*}{6w_1 \alpha_1^* - 1} \qquad (4\text{-}73)$$

4.6 AN AUGMENTED LAGRANGIAN

which is greater than zero provided w_1 is greater than $1/6\alpha_1^*$; again, this checks with a previous result, inequality (4-18).

4.7 MULTIPLIER UPDATE RULE VIA LOCAL DUALITY

In Example 4.3, an analytical expression for the local dual function $D(\eta)$ was obtained with relative ease because of the simple nature of the problem. For most practical problems, however, analytical solutions for $D(\eta)$ are not possible, and numerical techniques must be employed. Furthermore, complete knowledge of $D(\eta)$ for all η is not our primary objective; we would be satisfied to know η^* and $D(\eta^*)$.

Consider the following simplified procedure:

1) Holding the vector η fixed at η^k and the weights w_i's fixed at w^k, generate an unconstrained maximum of L_a with respect to x, and thereby obtain $x(\eta^k)$.

2) Use information concerning $dD(\eta^k)/d\eta$ and $d^2D(\eta^k)/d\eta^2$ to generate η^{k+1}, a new estimate of η^*.

3) Test the norm $||\eta^{k+1} - \eta^k||$: if this is sufficiently small, assume $\eta^{k+1} \simeq \eta^*$ and

stop; but if $\|\eta^{k+1} - \eta^k\|$ is too large, update w_i's to w^{k+1}, and return to Step 1 with k replaced by k + 1.

Our concern in the remainder of this section is a logical way in which to implement Step 2 above. Our approach parallels that in [L7]. From (4-55),

$$\nabla D(\eta^k) = -s(x(\eta^k)) \triangleq -s^k \qquad (4-74)$$

and from (4-61), using $\nabla s^k \triangleq \nabla s(x(\eta^k))$,

$$\nabla^2 D(\eta^k) = -\nabla s^k [\nabla^2 L_r(x(\eta^k), \eta^k)]^{-1} \nabla s^{k\prime}$$

$$= -\nabla s^k [\nabla^2 L_r^k]^{-1} \nabla s^{k\prime} \qquad (4-75)$$

The availability of (4-74) and (4-75) suggests the use of a one-step Newton procedure (Appendix E) to generate η^{k+1}, the next estimate of η^*. Thus,

$$\eta^{k+1} = \eta^k - [\nabla^2 D(\eta^k)]^{-1} \nabla D(\eta^k) \qquad (4-76)$$

and in terms of (4-74), (4-75), and (4-59),

$$\eta^{k+1} = \eta^k - 2w_1^k \left[\nabla s^k [\Psi \right.$$
$$\left. - \sum_{i=1}^{m_1+r} \nabla s_i^k \nabla s_i^{k\prime}]^{-1} \nabla s^{k\prime} \right]^{-1} s^k \qquad (4-77)$$

4.7 AN AUGMENTED LAGRANGIAN

in which the matrix Ψ is defined by

$$\Psi \triangleq [\nabla^2 f - \sum_{i=1}^{m_1+r} (\eta_i^k + 2w_1^k s_i^k) \nabla^2 s_i^k]/2w_1^k \quad (4\text{-}78)$$

Now, we would like to avoid having to evaluate the second derivative terms contained in (4-78); and fortunately, the larger w_1^k is made, the less influence these terms have in (4-77). For sufficiently large w_1^k, Ψ is called a <u>residue matrix</u> and is important in (4-77) only to the extent that it insures the existence of the inverse. In (4-77),

$$\nabla s^k [\Psi - \sum_{i=1}^{m_1+r} \nabla s_i^k \nabla s_i^{k\prime}]^{-1} \nabla s^{k\prime} \approx -I \quad (4\text{-}79)$$

when $w_1^k \gg 1$, with the result that (4-77) can be approximated by

$$\eta^{k+1} = \eta^k + 2w_1^k s^k \quad (4\text{-}80)$$

for sufficiently large w_1^k.

A component of (4-80) that applies for a typical equality constraint is

$$\alpha_i^{k+1} = \alpha_i^k + 2w_1^k [p_i(x^k) - a_i] \quad (4\text{-}81)$$

Note that (4-81) is equivalent to equation (4-38) which is the previously derived multiplier update rule for equality constraints. For active inequality constraints, (4-80) applies if the appropriate components of η^k are close to corresponding ones of η^{k+1}, but if (4-80) suggests large changes in the β_j's, such that a given β_j tends to be negative, we continue to restrict all β_j's ≥ 0, and therefore use (4-39) and (4-40) as appropriate update rules for β_j's.

4.8 CONVERGENCE RATES AS A FUNCTION OF WEIGHTS

From equations (4-74), (4-76), and (4-80), we observe that all of the eigenvalues of the Hessian for the dual problem are close to unity for large weights. The final convergence rate [L7] of the dual problem solution would therefore appear to be enhanced by the use of large weights. On the other hand, if large weights are used in the solution of the primal problem, large differences generally exist between the largest and the smallest eigenvalues of the Hessian matrix, a property that is known to produce poor convergence rates

4.8 AN AUGMENTED LAGRANGIAN

(see Appendix E). Now, by the nature of the multiplier algorithms, the sequence of weights specified for the primal problem solution are the same as that used in the dual problem solution, the two solutions being generated repetitively in an alternating sequence. It would appear, therefore, that a compromise must be made in the selection of weights. When an effective multiplier algorithm is applied to a particular problem starting with an initial (x,η) pair, there exists an optimum sequence of weights that result in $(x^k,\eta^k) \to (x^*,\eta^*)$ at a maximum terminal rate of convergence.

As noted previously, the weights must be large enough initially to avoid constraint breakthroughs during any unconstrained maximization phase, but not so large that the solution path is unduly confined during initial iterations of the algorithm when x is remote from x^*. Inefficient optimization may result if too much emphasis is placed on satisfying nonlinear constraints over the entire solution path (see, for example, the Around-the-World problem, test problem T7 of Chapter 6).

Another reason for having the weights increase as the solution progresses is that, for some of those constrained maxima which are abnormal or which do not satisfy second-order sufficient conditions, the final w_i's must be large to obtain an accurate approximation to the solution.

4.9 SUMMARY

In this chapter we have presented an unconstrained function L_a, that is formulated by augmenting the Lagrangian with weighted problem functions of the constrained NLP. The relationships between normal NLP solutions and the unconstrained maxima of L_a are established by Theorems 4.1 through 4.4. When second-order sufficient conditions are satisfied by a constrained maximum, finite penalty terms can be used in L_a. When second-order sufficient conditions are not satisfied, however, the constrained maximum can yet be obtained by operating on L_a with increasingly large penalty weights.

The major features of a multiplier algorithm are presented including the general procedure for updating the multipliers following each

4.9 AN AUGMENTED LAGRANGIAN

unconstrained maximization phase such that, as the sequence of maximizers $\{x^m\}$ converges to a constrained local maximum x^*, the multipliers converge to their optimal values α^* and β^*. Local duality theory is presented; it gives additional insight into the multiplier-update rule that is used in the algorithm. When applied to a general class of linear programming problems, the algorithm converges in a finite number of iterations. Details concerning the practical implementation of the algorithm are given in the next chapter.

PROBLEMS

4.1 Given $f = x_2 - x_1^2$ and $p = x_1^2 + x_2^2 = 1$, form the Lagrangian L. Treat the multiplier α as a fixed parameter and find the range of α over which L exhibits an unconstrained maximum. Is there a range of α over which L exhibits an unconstrained minimum? What is the true constrained maximum of f? (A problem of this type is considered by Forsythe [F12].)

4.2 Express the Hessian $\nabla^2 L_a$ by appropriate differentiation of (4-3). Note the discontinuity of $\nabla^2 L_a$ if C_b is not empty.

4.3 Given to be maximized is $-(x_1 - 1)^2 - (x_2 - 2)^2$ subject to $x_1 + x_2 = 2$. Form L_a and find the maximum of L_a with respect to x. Over what range of w_1 does a maximum of this L_a exist?

4.4 Form the L_a that applies to Problem 2.5a. The locked-on constraints at x^* are $x_1 \leq 2$ and $x_1 + x_2 \leq 3$. Using only these active constraints at x^*, find the maximum of L_a with respect to x. Over what range of w_2 does a maximum of this L_a exist?

4.5 To illustrate Lemma 4.1, let $B = [1 \ 0 \ 0]'$, $r_1 = [0 \ 1 \ 2]'$, and let A equal a 3×3 diagonal matrix with diagonal entries 2, -1, and -1, respectively.

    a. Show graphically the region V of Lemma 4.1. Is $\Delta x'(-A)\Delta x > 0$ for $\Delta x \in V$?

    b. Label the distinct regions associated with δ_1 as defined by (4-27).

    c. Form F of (4-28). Over what range of λ is $\Delta x'(-F)\Delta x > 0$ for all Δx?

4.6 Verify the three stationary points given for L_a of (4-23).

4.7 Consider the maximization of $f = x_1 - x_1^2 - x_2^4$ subject to $p(x_1, x_2) = x_1 x_2 - x_1 = 0$.

    a. Form the Lagrangian L and show that first-order conditions are satisfied at $x = (0, 0)$.

    b. Are second-order sufficient conditions satisfied at $x = (0, 0)$? Second-order necessary conditions?

c. Show that a minimum of the augmented Lagrangian L_a approaches (0, 0) as $w \to \infty$.

(You may wish to use the LPNLP computer program to do this. See, for example, the results of test problem T17 in Chapter 6.)

4.8 Give a logical flow diagram for the general procedure described in Section 4.4.

4.9 Form the dual function that applies to Problem 4.3, but limit it to the case that $w_1 = 0$. (Why is $w_1 = 0$ valid for this problem?) Show that the minimum of the dual equals the constrained maximum of f.

4.10 Form the local dual function that applies to Problem 4.4 and obtain the optimal multipliers by minimizing this dual function.

4.11 Consider the problem of maximizing $f = x_1 - (x_1 + 2x_2)^2$ subject to the constraint $p_1(x) = x_2 - 2x_1 = 1$. Using the augmented Lagrangian $L_a = x_1 - (x_1 + 2x_2)^2 - \eta_1(x_2 - 2x_1 - 1) - w(x_2 - 2x_1 - 1)^2$, show that the dual function $D(\eta_1)$ is $D(\eta_1) = (-0.39w + 0.08 + 0.2\eta_1 - \eta_1\lambda - \lambda^2)/w$ where $\lambda \triangleq -0.2 - 0.5\eta_1$. What value of η_1 minimizes $D(\eta_1)$, and for what range of w does $D(\eta_1)$ exhibit a minimum

4.12 Solve Problem 2.24 using the augmented Lagrangian of this chapter.

4.13 Consider the following problem from [B5]. Minimize $f = x_1$ subject to $(x_2 - 2) - (3 - x_1)^3 \leq 0$ and $-(x_2 - 2) - (3 - x_1)^3 \leq 0$. Solve using the Lagrangian and verify graphically. What can be said about the behavior of this problem? Also, solve using the augmented Lagrangian of this chapter.

4.14 Appendix C contains a list of typical modified Lagrangians. From the list in Appendix C, form a modified Lagrangian that accounts for both equality and inequality constraints, and examine the properties that apply in the neighborhood of a constrained local maximum.

4.15 In Section 4.6, $L_r(x^*,\eta)$ equals $f(x^*)$ for any η because $s(x^*) = 0$. With this information, replace the right-hand inequality in (4-64) by a more restrictive relationship. Also, describe the local relationship between $D(\eta) = L_r(x(\eta),\eta)$ and $L_r(x^*,\eta)$.

5
IMPLEMENTATION OF THE ALGORITHM

5.1 INTRODUCTION

The multiplier algorithm solves the constrained maximization problem by a sequence of unconstrained maximization stages. The essential structure of this type of algorithm is shown in Figure 5-1. After initialization it consists of the iterative sequence: 1) at the current search point x^k, generate the search direction r^k from information available; 2) find the point $x^k + \rho r^k$ which yields the maximum of the augmented function in the r^k direction from x^k (ρ is a scaler); 3) calculate and update the status at the new search point x^{k+1}; 4) test conditions for updating multipliers, penalty weights, and other algorithm parameters, and perform the update if conditions are

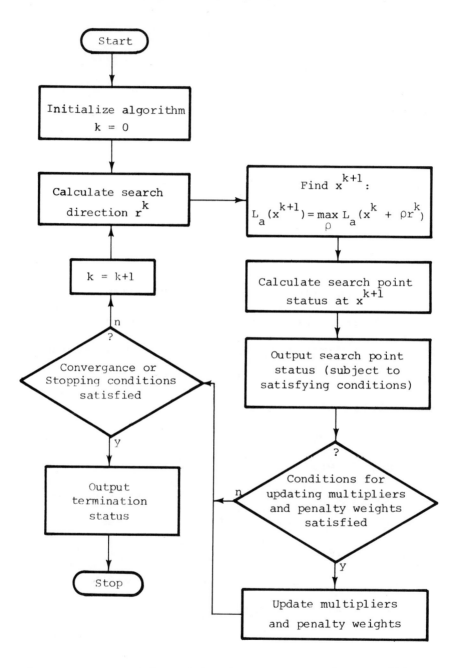

Figure 5-1   General maximization sequence using a multiplier algorithm.

5.1 IMPLEMENTATION OF THE ALGORITHM 177

satisfied; 5) test for convergence after every update phase, check stopping conditions in any case, and output terminal status if the search is to be terminated; and 6) if the process has not been terminated, repeat the steps from the new current search point x^{k+1}.

A FORTRAN code is given in Appendix B for solving the NLP by a comprehensive implementation of the above sequential process. Figure 5-2 shows the subroutines used to implement the multiplier algorithm. The general part of the code consists of 12 subroutines which are shown within the dashed lines; the code for these subroutines remains the same for every problem. Only the 3 remaining routines and data block require modification from problem to problem as they supply the analytical representation of a particular problem and information dictating operating modes of the general routine for that problem. An explanation of each subroutine, to describe its particular function in relation to the total algorithm, will now be given.

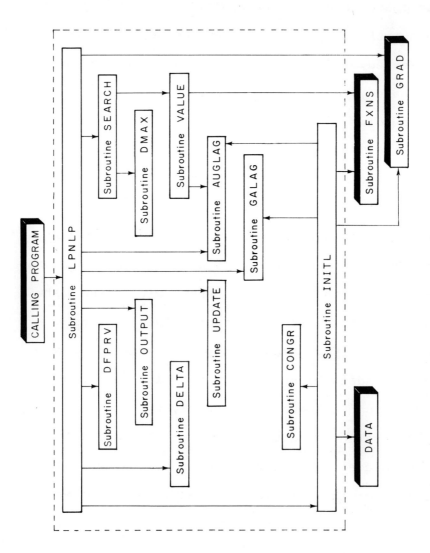

Figure 5-2 Subroutines used to implement the multiplier algorithm.

5.2 IMPLEMENTATION OF THE ALGORITHM

5.2 USER SUPPLIED SUBROUTINES

Three subroutines for any given problem and one data deck are required to supply the analytical model and parameter settings for the general routine. The short calling program properly dimensions all of the arrays and makes a call to the general working subroutine LPNLP. Subroutine LPNLP in turn directs subroutine INITL to read initial conditions and values for flags, counters, and parameters supplied in DATA, and to set operational modes of the algorithm. Once the algorithm is initialized, it performs the problem maximization sequentially, calling subroutines FXNS and GRAD whenever an evaluation of the problem model is required.

Subroutine FXNS assigns the objective function, $f(x)$, and the equality and inequality constraint functions, $p_i(x)$ ($i = 1, 2, \ldots, m_1$) and $q_j(x)$ ($j = 1, 2, \ldots, m_2$), respectively. The _dynamic_ gradient components (those dependent upon x) of the problem functions are assigned in subroutine GRAD. The constants a_i's of the equality constraints, the constants b_j's of the inequality constraints, and the lower bounds c_k's and upper bounds d_k's on the x_k's are assigned through the data base along with all nonzero <u>constant</u> gradient components of the problem functions. Appendix A contains an analytical model for the general NLP problem and detailed information required for dimensioning vectors, calling LPNLP, and supplying the data.

5.3 THE GENERAL ROUTINE LPNLP

In this section, the subroutines contained inside the dashed lines of Figure 5-2 are explained. A program listing for the general routine LPNLP is given in Appendix B. A brief explanation of each subroutine in the general routine follows.

SUBROUTINE LPNLP. Subroutine LPNLP is the organization and command program. It directs the maximization sequence as depicted in Figure 5-1. Control is given to this subroutine from the calling program. It then coordinates the operation of all remaining subroutines until the algorithm is

5.3 IMPLEMENTATION OF THE ALGORITHM

terminated, returning control to the calling program which then ends execution.

The flow connection for subroutine LPNLP is given in Figure 5-3. The terminology used in Figure 5-3 is described in Table 5-1. The algorithm is initialized and the operating modes are established by a call to INITL. After the initial starting-point function values are saved, the search direction r (R) is generated by DFPRV. Invariably,[†] LPNLP then calls subroutine SEARCH to find the value ρ_s (RHO) of ρ that gives $L_a(x+\rho_s r) \cong \max_\rho L_a(x+\rho r)$. If $\rho_s \neq 0$ following point C in Figure 5-3, the gradient (GAL) of L_a is evaluated by a call to GRAD and GALAG, and the search status is updated by DELTA. Appropriate counters are then incremented, and the search status is printed if NPRINT searches have been made since the last output or if $\rho_s = 0$.

[†] In unusual cases, the test associated with NG on the first page of Figure 5-3 may detect that DFPRV has returned five consecutive <u>gradient</u> search directions. Program execution is then terminated because such consecutive gradient searches generally yield a poor rate of convergence. The condition can also be caused by inconsistent function assignments in subroutines FXNS and GRAD.

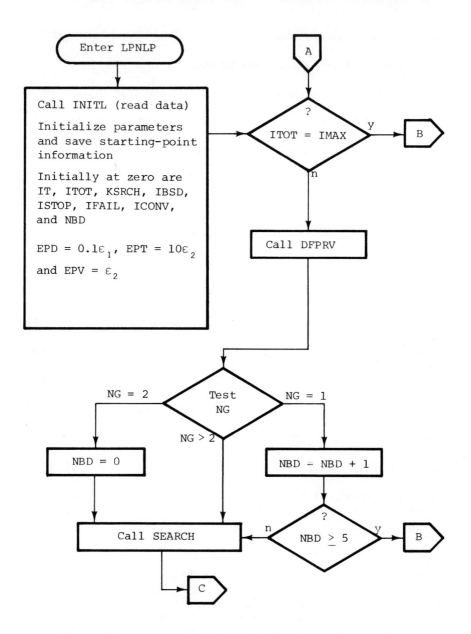

Figure 5-3 Flow connection for subroutine LPNLP.

5.3 IMPLEMENTATION OF THE ALGORITHM

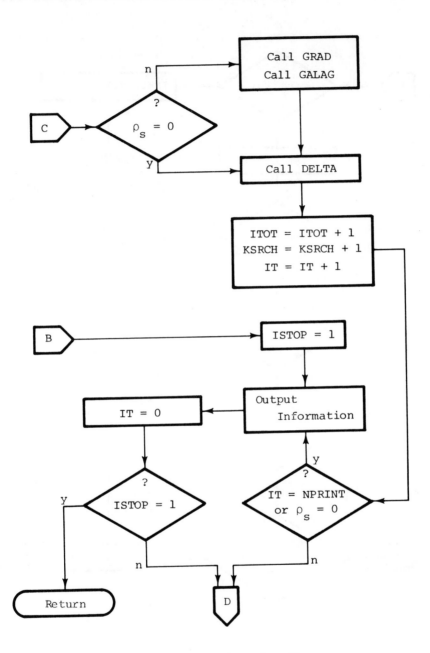

Figure 5-3 (continued).

184 MATH PROGRAMMING VIA AUGMENTED LAGRANGIANS

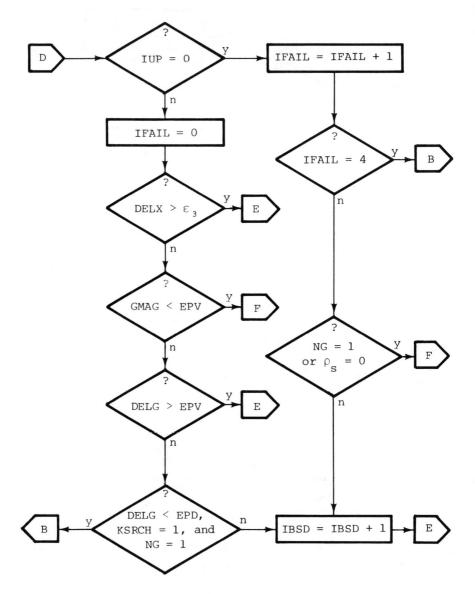

Figure 5-3 (continued).

5.3 IMPLEMENTATION OF THE ALGORITHM

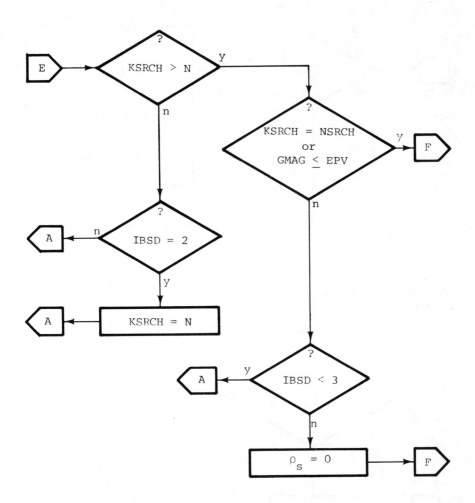

Figure 5-3 (continued).

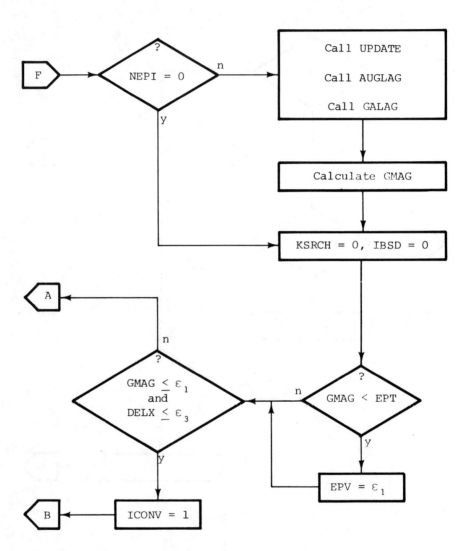

Figure 5-3 (continued).

5.3 IMPLEMENTATION OF THE ALGORITHM

Table 5-1

Notation used in Figure 5-3.

Name	Description
DELX	Magnitude of the change in x: $\|x^{k+1} - x^k\|$
DELG	Magnitude of the change in the gradient of L_a: $\|\nabla L_a^{k+1} - \nabla L_a^k\|$
ε_1[†]	Used in conjunction with GMAG (GMAG < ε_1) in a test for final convergence.
ε_2[†]	Initial value of EPV. ε_2 should be assigned a value larger than ε_1.
ε_3[†]	Used in conjunction with DELX (DELX < ε_3) both in a test to determine if multiplier updates are required and in a test for final convergence.
EPD	$0.1\varepsilon_1$
EPT	$10\varepsilon_2$
EPV	Initially assigned the value of ε_2, but is reassigned the value of ε_1 when GMAG following a multiplier update is less than EPT. Used in a set of tests to determine if a multiplier update is required.
GMAG	$\|\nabla L_a^k\|$

[†] Supplied by user in the data deck.

Table 5-1 (continued)

Name	Description
IBSD	Bad search direction counter. Set to zero initially and after any update of multipliers.
ICONV	Set to 0 initially; set to 1 when convergence conditions are satisfied.
IFAIL	Set to 0 initially and when subroutine SEARCH finds a larger value of L_a; incremented by 1 when a complete unidirectional search fails to find a larger value; IFAIL $\geq$ 4 terminates the algorithm.
IMAX[†]	Upper bound on the number of unidirectional searches allowed.
ISTOP	Set to 0 initially; set to 1 when program execution is to be terminated.
IT	Number of calls to subroutine SEARCH since the last call to subroutine OUTPUT.
ITOT	Total number of calls to subroutine SEARCH.
IUP	Set to 0 at the start of subroutine SEARCH; IUP = 1 is assigned in subroutine VALUE if SEARCH finds a larger value of L_a.

[†] Supplied by user in the data deck.

5.3 IMPLEMENTATION OF THE ALGORITHM

Table 5-1 (continued)

Name	Description
KSRCH	Normally, the number of calls to SEARCH since the last update of multipliers. However, KSRCH is assigned the value N if IBSD = 2 following point E of Figure 5-3.
N[†]	The number of x_i's.
NBD	Depends on NG: NBD = 0 if NG > 1; NBD = NBD + 1 if NG = 1.
NEPI	The number N, plus the number of equality constraints, plus the number of inequality constraints, plus the number of bounded variables.
NG	Set to 1 in DFPRV if the search direction is the gradient direction; otherwise, NG is incremented by 1 in DFPRV.
NPRINT[†]	Subroutine output is called when IT equals NPRINT.
NSRCH[†]	A multiplier update is forced when KSRCH reaches the bound NSRCH.
ρ_s (RHO)	The value of ρ returned by SEARCH; $x^{k+1} = x^k + \rho_s r^k$.

[†] Supplied by user in the data deck.

Tests to determine if the last unidirectional search made a significant change are conducted following point D in Figure 5-3. One of three possible results occurs: 1) the search is terminated by branching to point B; 2) the multiplier-update phase is called by branching to F; or 3) a branch is made to E from which additional tests are made to determine if a new search direction is to be generated by branching to A or if a multiplier-update phase is to be called by branching to F. With details contained in the flow diagram and in Table 5-1, we see that result number one above occurs when IUP = 0 (SEARCH failed to find a larger value of L_a) and IFAIL = 4 (four such failures in a row), or when IUP $\neq$ 0 but DELX $\leq$ ϵ_3, GMAG > EPV, DELG < EPD, KSRCH = 1, and NG = 1.

An update phase is called if possibility number one of the preceding paragraph does not apply and if one or more of the following sets of conditions are satisfied: 1) IUP = 0, IFAIL < 4, and NG = 1 (the past search direction was the gradient direction) or ρ_s = 0; 2) IUP $\neq$ 0, DELX < ϵ_3, and GMAG < EPV; 3) KSRCH > N and GMAG $\leq$ EPV; 4) KSRCH = NSRCH > N; or 5) KSRCH > N and IBSD $\geq$ 3.

5.3 IMPLEMENTATION OF THE ALGORITHM

At point F, if at least one constraint applies to the problem being solved (if NEPI $\neq$ 0), the multipliers and weights are updated, and counters KSRCH and IBSD are reset to 0. Following the update, if GMAG is less than $10\varepsilon_2$, EPV is assigned the value of ε_1 for use in subsequent tests for multiplier updates. Convergence conditions GMAG $\leq \varepsilon_1$ and DELX $\leq \varepsilon_3$ are then checked: if satisfied, final results are printed; if not satisfied, program control returns to point A.

SUBROUTINE INITL. The purpose of this subroutine is to effect the initialization of the algorithm. Flags, counters, and some parameters are set by INITL, and others are read from the data deck (see Appendix A). In this subroutine, all gradient and multiplier vectors are set to zero before initialization. Thus, only nonzero components must be supplied either through the data deck or in subroutine GRAD. The initial x values are read; the penalty weights are set; the problem model is evaluated; and the augmented Lagrangian and its gradient are formed. Subroutine INITL then prints the initial conditions of all important variables, parameters, and flags before returning control to LPNLP.

SUBROUTINE CONGR. This subroutine is called by INITL. It outputs those constant gradient components which are initialized through the data deck.

SUBROUTINE DFPRV[†]. The search direction R is generated by this subroutine. A modified Davidon-Fletcher-Powell (DFP) conjugate gradient method [D2, F10] is used where the directions are generated according to

$$r^i = H^i g^i, \quad i = 0, 1, 2, \ldots \qquad (5-1)$$

where

$$H^0 = I \qquad (5-2)$$

and, with exceptions to be noted,

$$H^{i+1} = H^i - \frac{(\Delta x^i)(\Delta x^i)'}{(\Delta x^i)'(\Delta g^i)} - \frac{H^i(\Delta g^i)(\Delta g^i)'H^i}{(\Delta g^i)'H^i(\Delta g^i)},$$
$$i = 0, 1, 2, \ldots \qquad (5-3)$$

[†] The self-scale equations of DFPRV are described in Appendix E.

5.3 IMPLEMENTATION OF THE ALGORITHM

in which $\Delta x^i = x^{i+1} - x^i$, $\Delta g^i = g^{i+1} - g^i$, g^{i+1} is the gradient of $L_a(x)$ at the point x^{i+1}, and x^{i+1} is found from the relation

$$L_a(x^{i+1}) \cong \max_{\rho} L_a(x^i + \rho H^i g^i) \qquad (5-4)$$

The sequence $H^0, H^1, \ldots, H^{n-1}$ generated by (5-2) and (5-3) is a sequence of positive-definite matrices. As x approaches x^*, H^i tends to $(-A)^{-1}$, the inverse matrix of second partial derivatives of the quadratic approximation to $L_a(x)$ at x^*. Thus, the unidirectional search is conducted over $\rho > 0$ only. The importance of these properties is discussed in Appendix E.

A flow connection for this mode of the subroutine is given in Figure 5-4. Initially, the counter NG is incremented by 1, and the following conditions are tested:

    i) RHO = 0

    ii) KSRCH = 0 and IFAIL > 0

    iii) KSRCH = 0, IRESET = 1, and NG > N

If one or more of these conditions are satisfied, the search direction is reinitialized by setting

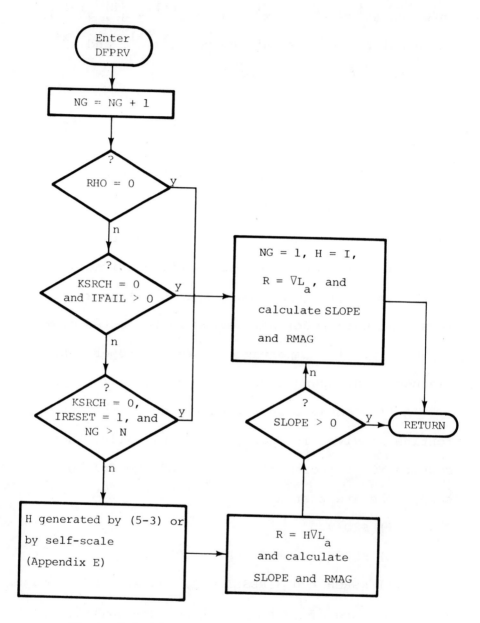

Figure 5-4 Flow connection for subroutine DFPRV.

5.3 IMPLEMENTATION OF THE ALGORITHM

the counter NG to 1, by setting H to I, as in equation (5-2), and by assigning the search direction to be the current gradient direction. If condition i is satisfied, the last unidirectional search failed to find a better point; if condition ii is satisfied, this is the first search of a new cycle[†] (KSRCH = 0) and the preceding unidirectional search failed to find a larger value of L_a (IFAIL > 0); if condition iii is satisfied, the automatic reset mode has been assigned by the user (IRESET = 1), the present search is the first of a new cycle (KSRCH = 0), and at least N unidirectional searches have been conducted since the last gradient-direction search. When not one of i, ii, or iii is satisfied, H is generated as in (5-3) (if ISS = 0), or H is generated by the use of related self-scale equations of Appendix E (if ISS = 1). Another reset condition is possible: if a denominator term of (5-3) is found to be zero because of numerical limitations, the reset block of Figure 5-4 is automatically employed.

[†] Each cycle generally consists of between N+1 and NSRCH (usually NSRCH = 2N + 1) unidirectional searches.

After H^i and r^i have been formed, one final test remains. A slope is calculated from (5-4) by

$$\text{SLOPE} = \left.\frac{dL_a}{d\rho}\right|_{\rho=0} = (r^i)'\nabla L_a(x^i) \qquad (5\text{-}5)$$

and RMAG is calculated by

$$\text{RMAG} = \|r^i\| = (r^{i\prime}r^i)^{\frac{1}{2}} \qquad (5\text{-}6)$$

The unidirectional search is conducted over positive stepsizes only, which is the case if the slope of (5-5) is positive. Therefore, if the computed SLOPE is less than or equal to zero, a reset to the gradient search direction is made, as is indicated in Figure 5-4. The generated r^i is returned as the direction for the next search.

SUBROUTINE SEARCH. This subroutine conducts a quadratic-convergent unidirectional search which is documented in Section 5.4. It seeks the stepsize ρ_s (RHO) that yields the maximum of $y(\rho)$,

$$y(\rho_s) = L_a(x^{i+1}) \cong \max_\rho L_a(x^i + \rho r^i) \qquad (5\text{-}7)$$

where $y(\rho)$ is a parametric representation of L_a.

5.3 IMPLEMENTATION OF THE ALGORITHM

SUBROUTINE VALUE. Subroutine VALUE is used by subroutine SEARCH in performing a unidirectional search. This subroutine evaluates the problem functions, by a call to FXNS, at a test point XT, which is a given stepsize D away from the initial point X of the unidirectional search, along the search direction R. The augmented Lagrangian L_a (AL) is formed by calling subroutine AUGLAG, and L_a is compared to the best value YS of the search. If L_a is a better value, it becomes the new best value of the search, and the stepsize D is stored along with the problem function values at XT.

SUBROUTINE DMAX. Subroutine DMAX estimates the maximum point of the local quadratic approximation of L_a by making a quadratic fit of 3 data values along the search direction. Depending upon the information available, either a point-slope-point fit is made, or a 3-point fit is made (as discussed in Section 5.4).

SUBROUTINE DELTA. Subroutine DELTA is called by subroutine LPNLP when SEARCH returns its best point. DELTA updates the search point status for the new point and calculates DELX, DELG, and GMAG which are used by LPNLP for testing convergence and updating criteria, and by DFPRV for generating the next search direction.

SUBROUTINE AUGLAG. This subroutine is called by INITL in initializing the algorithm, by SEARCH in conducting a unidirectional search, and by LPNLP after a multiplier and weight update. Its purpose is to formulate the augmented Lagrangian L_a (AL) at a given point. For this purpose, it uses the problem function values, the multipliers, and the weights at that point, as in equation (4-1).

SUBROUTINE GALAG. This subroutine is called by INITL in initializing the algorithm and by LPNLP after the best point is found by SEARCH. Its purpose is to formulate the gradient ∇L_a (GAL) of the augmented Lagrangian at this point from problem function values, problem function gradients, multipliers, and weights as in equation (4-3).

SUBROUTINE UPDATE. The purpose of this subroutine is to establish the constraint basis for each

5.3 IMPLEMENTATION OF THE ALGORITHM

search cycle by updating the multipliers according to the multiplier rules given by relations (4-38), (4-39), and (4-40). This subroutine also updates the penalty weights according to the weight update rule, relation (4-41).

SUBROUTINE OUTPUT. This subroutine prints the general search point information for the maximum of a unidirectional search. At the termination of the algorithm (ISTOP = 1), this subroutine prints the final values of the problem functions and the values of their associated multipliers.

5.4 THE UNIDIRECTIONAL SEARCH PROBLEM

In this section we discuss in detail the unidirectional quadratic-convergent search used in the multiplier algorithm.

If r is a specified direction vector in n-dimensional Euclidean space E^n,

$$x(\rho) = x^a + \rho r, \quad -\infty < \rho < \infty \quad (5-8)$$

is a line which passes through the point $x(0) = x^a$ in the line direction r. Consider $y(\rho)$ from (5-7),

$$y(\rho) = L_a(x^a + \rho r) \qquad (5-9)$$

The first derivative of $y(\rho)$ with respect to ρ is

$$\frac{\partial y}{\partial \rho} = r' \nabla L_a(x^a + \rho r) \qquad (5-10)$$

and the second derivative of $y(\rho)$ with respect to ρ is

$$\frac{\partial^2 y}{\partial \rho^2} = r' \nabla^2 L_a(x^a + \rho r) r \qquad (5-11)$$

Because of the $n(n + 1)/2$ functions that must be evaluated to obtain the symmetric matrix $\nabla^2 L_a$, and the additional operations that increase as n^3 to find $\nabla^2 L_a^{-1}$, as required by second-derivative search methods like the Newton search (Appendix E), it becomes quite evident that the more efficient n-dimensional search techniques make no direct use of second derivatives.

In any practical situation, the time spent in evaluating the model (i.e., objective and constraint functions and their gradients) at various points may well dominate the time for the whole maximization process. It is therefore desirable to make the unidirectional search as efficient as

5.4 IMPLEMENTATION OF THE ALGORITHM

possible using the generated data at previous points to the utmost in locating future points, and limiting the number of model evaluations as much as possible.

The unidirectional search problem then is to determine ρ_s such that

$$y(\rho_s) \cong \max_{\rho} y(\rho) = \max_{\rho} L_a(x^k + \rho r^k) \quad (5\text{-}12)$$

At $\rho = \rho_s$, $\dot{y}(\rho_s)$ satisfies†

$$\dot{y}(\rho_s) = \frac{\partial}{\partial \rho} y(\rho) \bigg|_{\rho=\rho_s} \cong 0 \quad (5\text{-}13)$$

In conducting a given unidirectional search, the value $y(0) = L_a(x^k)$ is usually available from the previous such search. Also assumed available is $\dot{y}(0) = (r^k)'\nabla L_a(x^k)$ where one aspect of r^k's selection is that it results in $\dot{y}(0) > 0$. It is only natural to use a search technique that utilizes the above information.

For the general constrained NLP, where the model is complex and the variables are highly

---

† In the remainder of the section, the derivative $\partial y / \partial \rho$ of a function y is denoted by $\dot{y}$.

interactive, it can be assumed that one gradient evaluation is approximately equivalent to n function evaluations. Thus, the required number of gradient evaluations in the unidirectional search should be held to a minimum. The search algorithm described in this section requires only one gradient evaluation per unidirectional search. This gradient evaluation occurs at the end of the search when ρ_s is located, and is used for generating the next search direction, for ensuring that $\partial y(0)/\partial \rho > 0$ at the beginning of the next search, and for testing the efficiency and overall convergence of the maximization process. In contrast, a single application of Davidon's cubic-convergent search [P4] may require three or more gradient evaluations per unidirectional search.

Two types of quadratic-convergent search are used in the algorithm. Consider the case where $y(\rho)$ is approximated locally by a quadratic of the form

$$y(\rho) = a(\rho - d_1)^2 + b(\rho - d_1) + c \qquad (5\text{-}14)$$

which, with $a < 0$, has a well-defined maximum at

5.4 IMPLEMENTATION OF THE ALGORITHM

$$\rho_s = d_1 - \frac{b}{2a} \qquad (5\text{-}15)$$

First, suppose two points, d_1 and d_2, are specified such that $y(d_1)$, $\dot{y}(d_1)$, and $y(d_2)$ can be evaluated. Since three independent data are sufficient to solve for all three constants of the quadratic approximation, the data $y(d_1)$, $\dot{y}(d_1)$, and $y(d_2)$, where $d_1 = 0$, are used in (5-14) to determine the constants a, b, and c such that ρ_s may be determined from (5-15). With a, b, and c determined by these three data, the maximum point ρ_s is

$$\rho_s = \frac{0.5(d_2)^2 \dot{y}(d_1)}{d_2 \dot{y}(d_1) + y(d_1) - y(d_2)} \qquad (5\text{-}16)$$

Suppose now a second situation where d_1, d_2, and d_3 are specified, and $y(d_1)$, $y(d_2)$, and $y(d_3)$ have been evaluated. Again, these three independent data are sufficient to determine a, b, and c from (5-14) such that the value of ρ_s can be found from (5-15). With a, b, and c determined by these three data, the maximum point ρ_s is

$$\rho_S = d_1 + 0.5 \frac{d_{21}^S y_{31} - d_{31}^S y_{21}}{d_{21} y_{31} - d_{31} y_{21}} \qquad (5-17)$$

where

$$\begin{aligned} y_{31} &= y(d_3) - y(d_1) \\ y_{21} &= y(d_2) - y(d_1) \\ d_{21} &= d_2 - d_1, \quad d_{31} = d_3 - d_1 \\ d_{21}^S &= (d_{21})^2, \quad d_{31}^S = (d_{31})^2 \end{aligned} \qquad (5-18)$$

Both (5-16) and (5-17) are woven into the following search technique as a means of estimating the maximum of $y(\rho)$. In examining the details of the algorithm, which is given in 17 phases, the following overall strategy should be held in view: Phases 1, 2, and 3 are initial phases of the algorithm. When necessary, Phases 4, 5, and 6 provide for large increases in the search parameter ρ. Phases 7 and 8 provide for significant decreases in ρ. Phases 9 and 10 allow moderate decreases in ρ; whereas Phases 11 and 12 allow moderate increases. Phases 14 through 17 are final phases. Note that Phases 1, 2, 3, and 17 are

5.4 IMPLEMENTATION OF THE ALGORITHM

involved in every unidirectional search; the other phases are used only as required.

<u>Phase 1</u>: Given are x^k, $L_a(x^k)$, $\nabla L_a(x^k)$, and r^k. Also, we have $y(\rho) = L_a(x^k + \rho r^k)$. Set $d_1 = 0$, and therefore $y(d_1) = L_a(x^k)$. Set flag IUP = 0, set counter IC1 = 0, and save initial best values. The search directions generated by LPNLP guarantee that the search can be conducted over $\rho > 0$ only; that is, $\dot{y}(0) > 0$.

<u>Phase 2</u>: The assigned value of the initial step d_2 depends on the following conditions.

    i) $10,000 < \|r^k\|$
    ii) $200 \leq \|r^k\| \leq 10,000$
    iii) KSRCH = 0 and IFAIL > 0 (see Table 5-1)
    iv) NG = 1 (indicating $r^k = \nabla L_a^k$)

$$d_2 = \begin{cases} 0.001 & \text{if i} \\ 10/\|r^k\| & \text{if ii} \\ 0.05 & \text{if iii but not i or ii} \\ 0.1 & \text{if iv but not i, ii, or iii} \\ 1.0 & \text{otherwise} \end{cases} \quad (5-19)$$

For most searches d_2 is assigned by $d_2 = 1$, but in the initial search iterations of the maximization process, the point x^k may be far away from the maximum x^*, and the magnitude of the search direction $||r^k||$ may be extremely large. When this situation arises, a value of 1 for d_2 may lead to a constraint breakthrough. Thus, for $||r^k|| > 200$, d_2 is modified according to (5-19).

Phase 3: With d_2 determined, $y(d_2)$ is evaluated, and the values $y(d_1)$, $\dot{y}(d_1)$, and $y(d_2)$ are used in (5-16) to locate the maximum point d_3 of the quadratic fit.[†] The value $y(d_2)$ is then compared to $y(d_1)$, with three possible results, as shown in Figure 5-5:

    a. If $y(d_2) > y(d_1)$, go to Phase 4.
    b. If $y(d_2) < y(d_1)$, go to Phase 7.
    c. If $y(d_2) = y(d_1)$, go to Phase 11.

When Phase 3c occurs, it is usually the result of d_2 being so close to d_1 that $y(d_2)$ cannot be distinguished from $y(d_1)$ within the numerical accuracy of the computer.

[†] If the denominator of (5-16) or (5-17) is computed to be zero, a nominal step $25d_2$ is assigned.

5.4 IMPLEMENTATION OF THE ALGORITHM 207

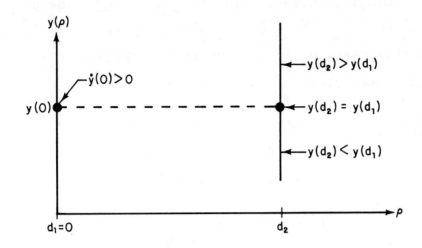

Figure 5-5 Possible situations considered in Phase 3 of unidirectional search.

Phase 4: $y(d_2) > y(d_1)$ where $d_1 = 0$

    a.   Set flag IUP = 1.

    b.   If $d_3 \leq 0$, go to Phase 11.

    c.   If $0.9d_2 \leq d_3 \leq 1.1d_2$, go to Phase 17.

    d.   Otherwise, continue with Phase 5.

If the condition of Phase 4b is satisfied, the quadratic fit of Phase 3 generated a minimum point, rather than a maximum point.[†] If the condition of 4c is satisfied, then d_3 is in the ±10% 'window' $w(d_2)$ of d_2, where

---

[†] This may occur when the algorithm is applied to a nonquadratic function in a region remote from x^*.

$$w(d_2) \triangleq [0.9d_2, 1.1d_2] \qquad (5-20)$$

In that case, d_2 is the accepted maximum point, and Phase 17, the final phase, is conducted.

<u>Phase 5</u>:

    a.   If $d_3 \leq 5.0d_2$, go to Phase 14.

    b.   If $d_3 > 100.0d_2$, reassign $d_3 = 100.0d_2$,

    c.   Evaluate $y(d_3)$.

    d.   If $y(d_3) > y(d_2)$, reassign $d_2 = d_3$, $y(d_2) = y(d_3)$, and go to Phase 6.

    e.   Otherwise, go to Phase 11.

Figure 5-6 illustrates the condition in Phase 5a.

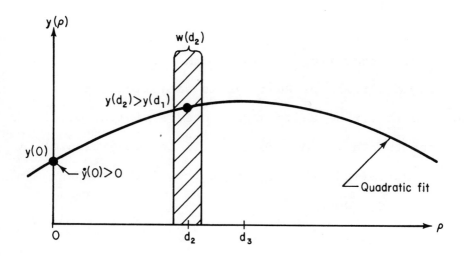

Figure 5-6   Representative case of Phase 5a.

5.4 IMPLEMENTATION OF THE ALGORITHM

Phase 6:

a. Use $y(d_1)$, $\dot{y}(d_1)$, and $y(d_2)$ in (5-16) to obtain d_3.

b. If $0.9d_2 \leq d_3 \leq 1.1d_2$, to to Phase 17.

c. If $d_3 \leq 0$, go to Phase 11.

d. If $IC1 \geq 10$, go to Phase 17.

e. Assign $IC1 = IC1 + 1$.

f. Go to Phase 5.

Phase 7:

a. If $d_3 > 0.2d_2$, go to Phase 9.

b. If $d_3 < 0.01d_2$, reassign $d_3 = 0.01d_2$.

c. Evaluate $y(d_3)$.

d. Reassign $d_2 = d_3$, $y(d_2) = y(d_3)$.

e. If $y(d_2) > y(d_1)$, assign $IUP = 1$, and go to Phase 6.

f. If $y(d_2) = y(d_1)$, go to Phase 11.

g. But if $y(d_2) < y(d_1)$, continue with Phase 8.

Figure 5-7 illustrates the condition in Phase 7a.

Phase 8:

a. If $IC1 \geq 10$, go to Phase 17.

b. Assign $IC1 = IC1 + 1$.

c. Use $y(d_1)$, $\dot{y}(d_1)$, and $y(d_2)$ in (5-16) to obtain d_3.

d. Return to Phase 7.

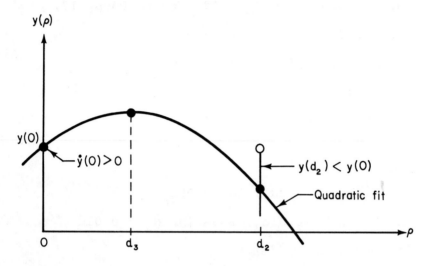

Figure 5-7 Representative case of Phase 7a.

Phase 9:

    a. Evaluate $y(d_3)$.

    b. If $y(d_3) > y(d_1)$, go to Phase 15.

    c. If $y(d_3) = y(d_1)$, go to Phase 17.

    d. If $y(d_3) < y(d_1)$, continue with Phase 10.

Phase 10:

    a. Reassign $d_2 = 0.2 d_3$.

    b. Assign IC1 = IC1 + 1.

    c. Evaluate $y(d_2)$.

5.4 IMPLEMENTATION OF THE ALGORITHM   211

    d.   If $y(d_2) = y(d_1)$, go to Phase 17.

    e.   If $y(d_2) > y(d_1)$ or if $IC1 > 10$, go to Phase 13.

    f.   But if $y(d_2) < y(d_1)$, reassign $d_3 = d_2$, $y(d_3) = y(d_2)$, and repeat Phase 10.

Phase 11:

    a.   Assign $d_3 = 5.0 d_2$.

    b.   Assign $IC1 = IC1 + 1$.

    c.   Evaluate $y(d_3)$.

    d.   If $y(d_3) > y(d_2)$, assign $IUP = 1$, and go to Phase 12.

    e.   If $y(d_3) = y(d_2)$, go to Phase 12.

    f.   If $y(d_3) < y(d_2)$ and $IUP = 0$, go to Phase 17.

    g.   But if $y(d_3) < y(d_2)$ and $IUP = 1$, go to Phase 13.

Figure 5-8 illustrates some of the conditions of Phases 11, 12, and 13.

Phase 12:

    a.   If $IC1 > 10$, go to Phase 13.

    b.   Otherwise, reassign $d_1 = d_2$, $y(d_1) = y(d_2)$, $d_2 = d_3$, $y(d_2) = y(d_3)$, and repeat Phase 11.

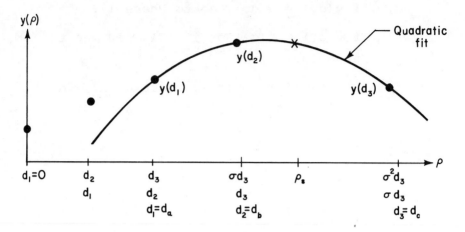

Figure 5-8 Representative case of Phases 11, 12 and 13 showing acceleration of points d_1, d_2, and d_3 away from $\rho = 0$. In the algorithm, $\sigma = 5$.

Phase 13:

    a.   Use (5-17) to obtain ρ_s.

    b.   Go to Phase 16.

Phase 14:

    a.   Evaluate $y(d_3)$.

    b.   If $y(d_3) \geq y(d_2)$, go to Phase 15.

    c.   Otherwise, go to Phase 13.

Phase 15:

    a.   Use (5-17) to obtain ρ_s.

    b.   If $0.9d_3 \leq \rho_s \leq 1.1d_3$, go to Phase 17.

    c.   Otherwise, continue with Phase 16.

5.4 IMPLEMENTATION OF THE ALGORITHM

<u>Phase 16</u>: Evaluate $y(\rho_s)$ and continue with Phase 17.

<u>Phase 17</u>: Assign RHO to DS (the best value of ρ), and assign other values of interest to those corresponding to the best value of ρ.

In order to effect Phase 17 above, the following is implicit in the preceding phases. Initially, $y(0)$ is saved as the best value y_{best} of the search, and DS = 0 is the initial best step. For each value of ρ at which $y(\rho)$ is evaluated, $y(\rho)$ is compared to y_{best}: if $y(\rho) \geq y_{best}$, DS is assigned the value of ρ, and y_{best} is assigned the value of $y(\rho)$.

In several situations in this search, tests are made to see if an estimated maximum point is within ±10% of a point already evaluated. If it is, the previously evaluated point is returned. This reduces the number of model evaluations for each unidirectional search. Whenever a point is evaluated, it is compared to the best point of the search. If this new point has a larger value of L_a, it becomes the best point. Since the maximum point of the last search is saved initially, we are guaranteed that the maximum of each search is greater than or equal to the value obtained in the previous search. Thus, the sequence $\cdots L_a(x^{k-1})$, $L_a(x^k)$, $L_a(x^{k+1})\cdots$ is a monotonic increasing sequence within each cycle. At the completion of each cycle, the multipliers and weights are updated. If the searches of each cycle have been effective, in that multipliers are converging to their optimum values, and if there is at least one value of L_a in each cycle that is larger than the largest value of L_a in the previous cycle, the sequence $\{L_a(x^k)\}$ tends to a constrained maximum of the problem as the sequence $\{x^k\}$ tends to x^*.

5.5 CONCLUSION

When implementing any algorithm on a computer, attention should be paid to making the algorithm convenient and versatile for the user. In this chapter, key features have been presented on an implementation of a multiplier algorithm for solving NLP's. Detailed attention was given to user-convenience and stopping criteria, with the result that the algorithm is not only invariably effective, but is also convenient for the user, allowing

5.5 IMPLEMENTATION OF THE ALGORITHM

both the terminating conditions and the amount of output to be controlled.

The arrangement of the subroutines for the multiplier algorithm is particularly convenient: the only subroutines that are problem dependent are the two in which the NLP functions and their derivatives are assigned. To solve a set of NLP's in which the NLP functions are fixed, but in which the NLP a_i's, b_j's, c_k's, and d_k's vary from problem to problem, only the data deck needs to be changed for each solution, and data decks can be stacked for multiple solution runs.

While many unidirectional search procedures could be used with the multiplier algorithm, the number of function evaluations required can vary widely when different unidirectional search procedures are employed. In Section 5.4, one such procedure is presented; as is evidenced in the next two chapters, it provides a good compromise between accuracy of unidirectional search and number of function evaluations.

The algorithm can be tailored to meet particular problem needs by the selection of various algorithm parameters. The examples in the next two chapters provide insight into how these parameters influence overall performance.

PROBLEMS

5.1 Subroutines FXNS and GRAD of Figure 5-2 are problem dependent. Form them for any NLP of interest and give the corresponding data set. Use the notation and procedures described in Appendix A. (Note that lower and upper bounds on x_i's are taken into account in the data and are included automatically in the augmented Lagrangian. They should not be included as inequality constraints in FXNS and GRAD.)

5.2 Analyze the use of EPV in LPNLP and explain why and under what conditions it is reduced to ε_1. How does this approach tend to satisfy (4-34)?

5.3 How should the signs in the DFP equations (5-1) and (5-3) be modified for performing minimization, rather than maximization?

5.4 There are five sets of conditions which can cause the search direction to be reset to the gradient direction in DFPRV. Summarize and explain these conditions.

5.5 Two points x^a and x^b are given in E^n. Find a straight line of the form (5-8) which passes through both x^a and x^b (solve for r in terms of x^a and x^b).

5.6 Derive (5-15) and (5-16).

PROBLEMS   IMPLEMENTATION OF THE ALGORITHM   217

5.7   Derive (5-17) and (5-18).

5.8   Given are $y(d_1)$, $\dot{y}(d_1)$, $y(d_2)$, and $y(d_3)$ where: $\dot{y}(d_1) > 0$, $d_1 < d_2 < d_3$, and $y(d_3) < y(d_2)$. Fit the cubic $a(\rho - d_1)^3 + b(\rho - d_1)^2 + c(\rho - d_1) + y(d_1)$ to the given data and find the local maximum point of the cubic in terms of the data.

5.9   Repeat Problem 5.8 given $y(d_1)$, $\dot{y}(d_1) > 0$, $y(d_2)$, $\dot{y}(d_2) < 0$, $d_1 < d_2$. A compact representation of the local maximum point of the cubic form is given by Davidon [D2].

5.10   Draw a logic flow diagram of the unidirectional search algorithm of Section 5.4. (This is more complex than one might expect.)

6
TEST PROBLEMS

6.1 INTRODUCTION

In this chapter the multiplier algorithm LPNLP is applied to several standard test problems, and the solutions are compared to those obtained by other methods which solve the NLP. Execution time to convergence or termination of the algorithms is a good standard of comparison, but is a function of the computer language and the type of computer which is used for solution, both of which vary significantly. Therefore, the comparison is made on the basis of the number of function evaluations required to reduce $|f(x) - f(x^*)|$ to a desired degree of accuracy. The solutions obtained for the problems presented here are the results of double-precision calculations on the XDS Sigma-7 computer at Montana State University.

6.2 UNCONSTRAINED TEST PROBLEMS AND RESULTS

The following problems (T1-T6) are unconstrained test problems and are compared to results obtained by Oren [O4,O5]. The parameter settings for LPNLP for T1-T6 are: $\varepsilon_1 = \varepsilon_2 = 10^{-7}$ and NSRCH = 2N + 2. For T1 and T6, $\varepsilon_3 = 10^{-7}$; for T4 and T5, $\varepsilon_3 = 10^{-6}$; and for T2 and T3, $\varepsilon_3 = 10^{-5}$. Results are given for three different operating modes of LPNLP: the DFP mode, the DFP mode with resets to the gradient direction every NSRCH searches; and the self-scaling variable metric mode of Oren with $\phi = 1$ and $\theta = 0.5$. The test problems are described next.

T1: BANANA FUNCTION [O4]

$$f(x) = -\sum_{k=1}^{N-1} [100(x_{k+1} - x_k^2)^2 + (1 - x_k)^2]$$

Starting point: $x_0 = (-1.2, 1, -1.2, 1, -1.2, \ldots)$. The maximum point is at $x^* = (1, 1, 1, 1, \ldots)$ at which $f(x^*) = 0$. For N = 2, the function is that of Rosenbrock [R9]. This problem was run for N = 2, 6, 10, 16, 30, 50, and 100. For N > 2, local maxima exist, as noted in footnote 4 of Table 6-1.

T2: WOOD'S FUNCTION [C5]

$$f(x) = -[100(x_2 - x_1^2)^2 + (1 - x_1)^2$$
$$+ 90(x_4 - x_3^2)^2 + (1 - x_3)^2$$
$$+ 10.1(x_2 - 1)^2 + 10.1(x_4 - 1)^2$$
$$+ 19.8(x_2 - 1)(x_4 - 1)]$$

Starting point: $x_0 = (-3, -1, -3, -1)$.
The optimum point is at $x^* = (1, 1, 1, 1)$ at which $f(x^*) = 0$.

T3: POWELL'S FUNCTION [P12]

$$f(x) = -[(x_1 + 10x_2)^2 + 5(x_3 - x_4)^2$$
$$+ (x_2 - 2x_3)^4 + 10(x_1 - x_4)^4]$$

Starting point: $x_0 = (3, -1, 0, 1)$.
The optimum point is at $x^* = (0, 0, 0, 0)$ at which $f(x^*) = 0$.

6.2 TEST PROBLEMS

T4: QUARTIC [O5]

$$f(x) = -(x'Qx)^2$$

where $Q = \text{diag}(1, 2, 3, \ldots, N)$.
Starting point: $x_0 = (1, 1, 1, \ldots, 1)$.
The optimum point is at $x^* = (0, 0, 0, \ldots, 0)$ at which $f(x^*) = 0$. This problem was run for $N = 10$, 30, and 50.

T5: BIGGS' FUNCTION [B6], $N = 3$.

$$f(x) = -\sum_{i=1}^{10} (e^{-x_1 z_i} - x_3 e^{-x_2 z_i} - y_i)^2$$

where

$$y_i = e^{-z_i} - 5e^{-10 z_i} \quad \text{and}$$

$$z_i = (0.1)i, \quad i = 1, 2, \ldots, 10$$

Starting point: $x_0 = (1, 2, 1)$.
The optimum point is at $x^* = (1, 10, 5)$ at which $f(x^*) = 0$.

T6: BIGGS' FUNCTION [B6], N = 5

$$f(x) = -\sum_{i=1}^{11} (x_3 e^{-x_1 z_i} - x_4 e^{-x_2 z_i} + 3e^{-x_5 z_i} - y_i)^2$$

where

$$y_i = e^{-z_i} - 5e^{-10 z_i} + 3e^{-4 z_i} \quad \text{and}$$

$$z_i = (0.1)i, \quad i = 1, 2, \ldots, 11$$

Four local maxima for this problem are as follows: 1) x^a = (1, 10, 1, 5, 4) with $f(x^a) = 0$; 2) x^b = (1.7767, 16.124, -0.59421, 4.7072, 1.7767) with $f(x^b) = -26.5$; 3) x^c = (-0.45160, -0.45160, -0.76571, 0.76571, 0.063525) with ($f(x^c) = -0.21551$; and 4) x^d = (10, 1, -5, -1, 4) with $f(x^d) = 0$.

Point a above was obtained by the three modes of LPNLP listed in Table 6-1 using an initial x of (2, 2, 2, 2, 2). The suggested starting point of (1, 2, 1, 1, 1) given in [O5] resulted in the following: with LPNLP-DFP, convergence to point a was obtained in (IT, NF) = (359, 921); with LPNLP-DFP/RESET, convergence to point a in (193, 473); and with LPNLP-SS, convergence to point b

6.2 TEST PROBLEMS

in (39, 102). Point c was obtained by LPNLP-DFP in (27, 75) starting at x = (0, 0, 0, 0, 0). Whereas point d was obtained by LPNLP-SS in (80, 201) starting at x = (1, 1, 1, 1, 1).

RESULTS FOR UNCONSTRAINED PROBLEMS

The results for the unconstrained test problems are tabulated in Table 6-1. The results given in Table 6-1 indicate that for the three modes of search direction generation for LPNLP, the DFP modes may have some advantage for a small number of variables. However, as the number of variables increases, the self-scaling (SS) mode for LPNLP becomes increasingly better. The SS mode was better than DFP for all unconstrained cases when $N \geq 6$ variables. This behavior can be explained by the fact that DFP has property 1 mentioned in Appendix E, while SS has property 2. The argument given in Appendix E regarding the tradeoffs between the two properties is supported by the results. These results are in agreement with those reported by Oren [04].

Table 6-1

Function	N	expo	LPNLP-DFP[1] ACC IT	NF	CONV IT	NF	expo	LPNLP-DFP/Reset[1] ACC IT	NF	CONV IT	NF
T1	2	-9	43	123	46	126	-9	30	101	33	106
"	6	--	--	--	45[4]	121	--	--	--	51[4]	151
"	10	--	--	--	67[4]	168	--	--	--	87[4]	256
"	16	-9	234	605	238	609	-13	153	393	155	397
"	30	-11	276	683	280	689	-13	294	724	296	726
"	50	-15	485	1147	486	1149	-11	495	1163	499	1167
"	100	-17	1210	2816	1210	2816	-17	1156	2605	1156	2605
T2	4	-12	88	238	90	240	-12	67	206	69	208
T3	4	-10	32	93	48	142	-10	20	57	22	61
T4	10	-10	36	106	46	128	-10	48	134	74	211
"	30	-10	109	304	126	346	-10	106	290	147	411
"	50	-10	139	390	142	397	-10	127	355	161	449
T5	3	-11	12	25	14	27	-12	15	34	17	37
T6	5	-11	135	343	139	348	-10	59	146	68	167

Table 6-1 (continued)

Function	N	LPNLP-SS(1) ACC expo			LPNLP-SS(1) CONV			SSVM ACC	
		expo	IT	NF	IT	NF	expo	IT	NF
T1	2	-10	28	77	31	83	-10	25(2)	71
"	6	-10	54	129	59	136	-10	45(2)	118
"	10	-10	70	150	74	157	-10	60(2)	150
"	16	-12	84	181	88	188	-10	80(2)	200
"	30	-11	129	304	134	312	-9	259(3)	749
"	50	-11	200	441	206	447	-9	437(3)	1319
"	100	-15	380	816	382	819			
T2	4	-10	83	235	86	241	-10	60(3,5)	183
T3	4	-10	32	70	47	102	-10	42(2)	82
T4	10	-10	15	47	28	86	-10	10(2)	31
"	30	-10	23	65	36	104	-10	18(2)	61
"	50	-10	28	80	42	122	-9	29(3)	88
T5	3	-10	13	33	16	37	-10	12(2)	33
T6	5	-10	32	76	37	86	-10	91(6)	266

ACC (accuracy) -- conditions corresponding to a point at which $|f(x) - f(x^*)| < 10^{\text{expo}}$. CONV (convergence) -- convergence based on $\|\nabla f(x)\| < \varepsilon_1$ and $\|x^{k+1} - x^k\| < \varepsilon_3$; at convergence, $|f(x) - f(x^*)|$ was usually $< 10^{-20}$. N -- number of variables. IT -- number of iterations. NF -- number of function evaluations.

Table 6-1 (continued)

(1) Unidirectional search used is that of Section 5.4. Requires 1 gradient evaluation per iteration.

(2) Data from [O5] with $\delta = 0.5$, $\phi = 1$, $\theta = 0.5$. Unidirectional search used is that of cubic interpolation which requires ≥ 3 gradient evaluations per iteration if Davidon's method is used.

(3) Data from [O4]. Values listed are from convergence using $\phi = \theta = 0$ in SSVM. NF in this case is cumulative number of function and gradient evaluations.

(4) Converged to a local maximum: for N = 6, local x = (-0.986575, 0.983398, 0.972107, 0.947437, 0.898651, 0.807574) at which f = -3.97394 and $||\nabla f|| < 10^{-10}$. A similar point applies for N = 10.

(5) Oren's description [O4] of this problem uses "... 90$(x_4 - x_3^2)^2$ ..." instead of the original term "... 90$(x_4 - x_2^2)^2$ ..." [C5]. When LPNLP-SS is applied to Oren's version, the results are (IT, NF) = (39, 101) for ACC.

(6) The starting point for this case was (1, 2, 1, 1, 1). See the description of the T6 test problem.

6.2 TEST PROBLEMS

SSVM is tested in [O4] with restarts every $N + 1$ iterations. In practically all cases, SSVM with restarts every $N + 1$ iterations yielded poorer results that SSVM with no restarts. Table 6-1 shows that LPNLP/DFP does better without regular resets every $2N + 2$ iterations in 8 of 14 cases. The results from Table 6-1 indicate that LPNLP/SS is comparable to SSVM when SSVM uses $\phi = 1$, $\theta = 0.5$, and $\delta = 0.5$. The parameter δ in [O5] controls the frequency of unidirectional searching: with δ close to zero, only one evaluation of the model is generally made between each search direction generated; but when $\delta = 0.5$, a unidirectional search is made at each step. While smaller values of δ in [O5] generally gave smaller NF, they also produced larger values for IT and for overall execution time.

Based on the given results, LPNLP/SS appears to be the best mode of operation for the LPNLP algorithm when applied to unconstrained problems having 6 or more variables. For problems with $N < 6$, the DFP modes have the advantage in many instances.

6.3 CONSTRAINED TEST PROBLEMS AND RESULTS

Of the following constrained test problems, T7, T8 and T9 are considered in [P7], and T10 through T16 are used in [A8]. For each problem, the important parameter settings are listed (i.e., WF, NSRCH, ε_i, w_i and $w_i$max for i = 1, 2, 3). For all cases, the multiplier vectors α and β are initially set to zero. Results are given for four modes of the LPNLP algorithm. The test problems are described next.

T7: AROUND THE WORLD (Pierre [P7]). The problem is to find the maximum of f on a sphere, subject to remaining on one side of a plane that passes through the sphere (Figure 6-1).

Maximize

$$f = x_2$$

subject to

$$x_1^2 + x_2^2 + x_3^2 = 1, \text{ and}$$

$$2x_2 - x_1 \leq 1$$

6.3 TEST PROBLEMS   229

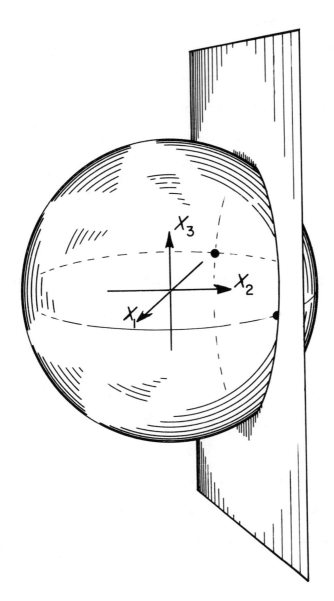

Figure 6-1   The geometry of problem T7.

Starting point: $x_0 = (-0.1, -1, 0.1)$, $\varepsilon_1 = 10^{-6}$, $\varepsilon_2 = \varepsilon_3 = 10^{-2}$, $w_i = 0.25$, $w_{i\max} = 1$, $i = 1, 2, 3$, WF = 2, and NSRCH = 7.

Computational results for this problem and others are given in Table 6-3. Not shown in Table 6-3, however, are the results obtained for this problem under several different conditions on the w_i's, but with everything else the same. Table 6-2 contains these limited results. Note that the larger weights caused too much emphasis to be given to remaining close to the sphere during the initial phases of the search. During the final phases of the search, however (say, from 4 place accuracy to convergence), the larger weights performed best.

The optimum point is at $x^* = (3/5, 4/5, 0)$ at which $f(x^*) = 4/5$, $\alpha^* = 1/4$ and $\beta^* = 3/10$.

A saddle point x_s exists at $x_s = (-1, 0, 0)$ at which $f(x_s) = 0$, $\alpha = 0.25$ and $\beta = 0.5$. First-order necessary conditions are satisfied at the saddle point, but both larger and smaller values of f are to be found in constrained neighborhoods of x_s. A slight perturbation (initial $x = (-1.0, 0, 0.05)$)

6.3 TEST PROBLEMS

Table 6-2

Function Evaluations as a Function of Weights

for Around-the-World (DFP mode)

Initial w_i	Accuracy in f $w_{i_{max}}$	expo = -3 (1)		expo = -4		expo = -5		Convergence	
		IT	NF	IT	NF	IT	NF	IT	NF
0.025	0.1	19	36	23	40	25	42	34	51
0.25	1.0	13	36	15	39	17	41	23	53
2.5	10.0	54	155	54	155	58	159	61	162
25.0	100.0	97	288	104	302	104	302	110	308

(1) - IT = Number of unidirectional searches.
NF = Number of calls to subroutine FXNS.
Results shown are those at which $|f(x) - f(x^*)| < 10^{expo}$.

away from the saddle point causes the LPNLP solution to converge to the optimum.

T8: A LINEAR PROBLEM (Pierre [P7]).

Maximize

$$f = 0.5x_1 + x_2 + 0.5x_3 + x_4$$

subject to

$$x_1 + x_2 + x_3 - 2x_4 = 6$$

$$\sum_{i=1}^{4} x_i \leq 10$$

$$0.2x_1 + 0.5x_2 + x_3 + 2x_4 \leq 10$$

$$2x_1 + x_2 + 0.5x_3 + 0.2x_4 \leq 10$$

$$x_i \geq 0, \ i = 1, 2, 3, 4$$

Starting point: $x_0 = (0, 0, 0, 0)$, $\varepsilon_1 = 10^{-6}$, $\varepsilon_2 = \varepsilon_3 = 10^{-2}$, $w_i = 1$, $w_{imax} = 16$, $i = 1, 2, 3$, WF = 2, NSRCH = 9.

The optimum point is at $x^* = (0, 26/3, 0, 4/3)$ at which $f(x^*) = 10$, $\alpha^* = 0$, and $\beta^* = (1, 0, 0, 0.5, 0, 0.5, 0)$ where the last four components of β are the multipliers associated with $x_i \geq 0$, $i = 1, 2, 3, 4$.

6.3 TEST PROBLEMS

T9: SEVEN VARIABLE PROBLEM (Pierre [P7]).

Maximize

$$f = 5x_1 + 5x_2 + 4x_3 + x_1 x_3 + 6x_4 + \frac{5x_5}{1 + x_5}$$

$$+ \frac{8x_6}{1 + x_6} + 10(1 - 2e^{-x_7} + e^{-2x_7})$$

subject to

$$2x_4 + x_5 + 0.8x_6 + x_7 = 5$$

$$x_2^2 + x_3^2 + x_5^2 + x_6^2 = 5$$

$$\sum_{i=1}^{7} x_i \le 10$$

$$\sum_{i=1}^{4} x_i \le 5$$

$$x_1 + x_3 + x_5 + x_6^2 - x_7^2 \le 5$$

$$x_i \ge 0, \quad i = 1, 2, \ldots, 7$$

Starting point: $x_{0i} = 0.1$, $i = 1, 2, \ldots, 7$, $\varepsilon_1 = 10^{-4}$, $\varepsilon_2 = \varepsilon_3 = 0.1$, $w_i = 1$, $w_{imax} = 32$, $i = 1, 2, 3$, WF = 4, and NSRCH = 15.

A constrained local maximum exists at $x^* =$ (3.24182, 0, 1.63416, 0.124021, 0.889614, 1.24021,

2.87018) at which $f(x^*) = 44.4687$, $\alpha^* = (-0.317079, 0.185926)$, and $\beta^* = (1.38658, 5.24758, 0, 0, 1.63416, 0, 0, 0, 0, 0)$, where the last seven components of β are the multipliers associated with $x_i \geq 0$, $i = 1, 2, \ldots, 7$.

T10: (suggested by Fiacco and McCormick [F4]).
Maximize

$$f = -\tfrac{1}{3}(x_1 + 1)^3 - x_2$$

subject to $x_1 \geq 1$ and $x_2 \geq 0$.
Starting point: $x_0 = (1.125, 0.125)$, $\varepsilon_1 = 10^{-6}$, $\varepsilon_2 = \varepsilon_3 = 10^{-2}$, $w_1 = w_{1max} = 0$, $w_2 = w_3 = 1$, $w_{2max} = w_{3max} = 64$, $WF = 4$, $NSRCH = 5$.

This function has an optimum point at $x^* = (1, 0)$ at which $f(x^*) = -8/3$, and $\beta^* = (4, 1)$.

T11: (Rosen and Suzuki's problem [R8]).
Maximize

$$f = 5(x_1 + x_2) + 7(3x_3 - x_4) - x_1^2 - x_2^2 - 2x_3^2 - x_4^2$$

subject to

6.3 TEST PROBLEMS

$$\left(\sum_{i=1}^{4} x_i^2\right) + x_1 - x_2 + x_3 - x_4 \leq 8$$

$$x_1^2 + 2x_2^2 + x_3^2 + 2x_4^2 - x_1 - x_4 \leq 10$$

$$2x_1^2 + x_2^2 + x_3^2 + 2x_1 - x_2 - x_4 \leq 5$$

Starting point: $x_0 = (0, 0, 0, 0)$, $\varepsilon_1 = 10^{-6}$, $\varepsilon_2 = \varepsilon_3 = 10^{-2}$, $w_1 = w_{1max} = 0$, $w_2 = w_3 = 1$, $w_{2max} = w_{3max} = 16$, WF = 4, NSRCH = 9.

A constrained local maximum exists at $x^* = (0, 1, 2, -1)$ at which $f(x^*) = 44$, and $\beta^* = (1, 0, 2)$.

T12: (Beale's Problem [B2]).

Maximize

$$f = 8x_1 + 6x_2 + 4x_3 - 2x_1^2 - 2x_2^2 - x_3^2 - 2x_1x_2 - 2x_1x_3 - 9$$

subject to

$$x_1 + x_2 + 2x_3 \leq 3, \quad \text{and}$$

$$x_i \geq 0, \quad i = 1, 2, 3$$

Starting point: $x_0 = (\tfrac{1}{2}, \tfrac{1}{2}, \tfrac{1}{2})$, $\varepsilon_1 = 10^{-6}$,

$\varepsilon_2 = \varepsilon_3 = 10^{-2}$, $w_1 = w_{1\max} = 0$, $w_2 = w_3 = 1$, $w_{2\max} = w_{3\max} = 16$, WF = 4, NSRCH = 7.

This function has a constrained local maximum at the point $x^* = (4/3, 7/9, 4/9)$ at which $f(x^*) = -1/9$, and $\beta^* = (2/9, 0, 0, 0)$, where the last three components of β are the multipliers associated with $x_i \geq 0$, $i = 1, 2, 3$.

T13: (suggested by Powell [P12]).

Maximize

$$f = -x_1 x_2 x_3 x_4 x_5$$

subject to

$$\sum_{i=1}^{5} x_i^2 = 10$$

$$x_2 x_3 - 5 x_4 x_5 = 0$$

$$x_1^3 + x_2^3 = -1$$

Starting point: $x_0 = (-2, 1.5, 2, -1, -1)$, $\varepsilon_1 = 10^{-6}$, $\varepsilon_2 = \varepsilon_3 = 10^{-2}$, $w_1 = 0.5$, $w_{1\max} = 8$, $w_i = w_{i\max} = 0$, $i = 2, 3$, WF = 2, NSRCH = 1.

This function has a constrained local maximum at $x^* = (-1.71714, 1.59571, 1.82725, -0.763643,$

6.3 TEST PROBLEMS

-0.763643) at which $f(x^*) = 2.91970$, and $\alpha^* = (0.744446, -0.703575, 0.096806)$.

Powell notes that some difficulty could occur with constraint breakthrough because f is not bounded from above. Therefore, $f = -\exp(x_1 x_2 x_3 x_4 x_5)$ is the recommended function to be maximized subject to the same constraints as above. No constraint breakthrough was encountered with the problem, and the results were obtained with the problem function as first stated.

T14: (suggested by Wong [W4], see also [A8]).

Maximize

$$f = -(x_1 - 10)^2 - 5(x_2 - 12)^2 - x_3^4$$
$$- 3(x_4 - 11)^2 - 10x_5^6 - 7x_6^2$$
$$- x_7^4 + 4x_6 x_7 + 10x_6 + 8x_7$$

subject to

$$2x_1^2 + 3x_2^4 + x_3 + 4x_4^2 + 5x_5 \leq 127$$
$$7x_1 + 3x_2 + 10x_3^2 + x_4 - x_5 \leq 282$$
$$23x_1 + x_2^2 + 6x_6^2 - 8x_7 \leq 196$$
$$4x_1^2 + x_2^2 - 3x_1 x_2 + 2x_3^2 + 5x_6 - 11x_7 \leq 0$$

Starting point: $x_0 = (1, 2, 0, 4, 0, 1, 1)$, $\varepsilon_1 = 10^{-4}$, $\varepsilon_2 = \varepsilon_3 = 0.1$, $w_1 = w_{1max} = 0$, $w_2 = w_3 = 0.25$, $w_{2max} = w_{3max} = 16$, WF = 4, NSRCH = 15

A constrained local maximum of this function exists at $x^* = (2.33050, 1.95137, -0.477541, 4.36573, -0.624487, 1.03813, 1.59423)$ at which $f(x^*) = -680.630$, and $\beta^* = (1.13972, 0, 0, 0.368615)$.

T15: (suggested by Wong [W4], see also [A8]).

Maximize

$$f = -x_1^2 - x_2^2 - x_1 x_2 + 14x_1 + 16x_2$$
$$- (x_3 - 10)^2 - 4(x_4 - 5)^2 - (x_5 - 3)^2$$
$$- 2(x_6 - 1)^2 - 5x_7^2 - 7(x_8 - 11)^2$$
$$- 2(x_9 - 10)^2 - (x_{10} - 7)^2 - 45$$

subject to

$$3(x_1 - 2)^2 + 4(x_2 - 3)^2 + 2x_3^2 - 7x_4 \le 120$$
$$5x_1^2 + 8x_2 + (x_3 - 6)^2 - 2x_4 \le 40$$
$$0.5(x_1 - 8)^2 + 2(x_2 - 4)^2 + 3x_5^2 - x_6 \le 30$$

6.3 TEST PROBLEMS

$$x_1^2 + 2(x_2 - 2)^2 - 2x_1x_2 + 14x_5 - 6x_6 \le 0$$

$$4x_1 + 5x_2 - 3x_7 + 9x_8 \le 105$$

$$10x_1 - 8x_2 - 17x_7 + 2x_8 \le 0$$

$$-3x_1 + 6x_2 + 12(x_9 - 8)^2 - 7x_{10} \le 0$$

$$-8x_1 + 2x_2 + 5x_9 - 2x_{10} \le 12$$

Starting point: x_0 = (2, 3, 5, 5, 1, 2, 7, 3, 6, 10), $\varepsilon_1 = 10^{-4}$, $\varepsilon_2 = \varepsilon_3 = 10^{-2}$, $w_1 = w_{1max} = 0$, $w_2 = w_3 = 0.25$, $w_{2max} = w_{3max} = 4$, WF = 4, NSRCH = 21.

This function has a constrained local maximum at x^* = (2.17200, 2.36368, 8.77393, 5.09598, 0.990655, 1.43057, 1.32164, 9.82873, 8.28009, 8.37593) at which $f(x^*) = -24.3062$, and β^* = (0.020545, 0.312029, 0, 0.287049, 1.71653, 0.474520, 0, 1.37593).

T16: (suggested by Wong [W4], see also [A8]).
Maximize $f = f_1 + f_2$ where

$$f_1 = -x_1^2 - x_2^2 - x_1x_2 + 14x_1 + 16x_2 - (x_3 - 10)^2 - 4(x_4 - 5)^2 - (x_5 - 3)^2$$

and

$$f_2 = -2(x_6 - 1)^2 - 5x_7^2 - 7(x_8 - 11)^2$$
$$- 2(x_9 - 10)^2 - (x_{10} - 7)^2$$
$$- (x_{11} - 9)^2 - 10(x_{12} - 1)^2$$
$$- 5(x_{13} - 7)^2 - 4(x_{14} - 14)^2$$
$$- 27(x_{15} - 1)^2 - x_{16}^4 - (x_{17} - 2)^2$$
$$- 13(x_{18} - 2)^2 - (x_{19} - 3)^2 - x_{20}^2 - 95$$

subject to

$$3(x_1 - 2)^2 + 4(x_2 - 3)^2 + 2x_3^2 - 7x_4 \le 120$$
$$5x_1^2 + 8x_2 + (x_3 - 6)^2 - 2x_4 \le 40$$
$$0.5(x_1 - 8)^2 + 2(x_2 - 4)^2 + 3x_5^2 - x_6 \le 30$$
$$x_1^2 + 2(x_2 - 2)^2 - 2x_1 x_2 + 14x_5 - 6x_6 \le 0$$
$$4x_1 + 5x_2 - 3x_7 + 9x_8 \le 105$$
$$10x_1 - 8x_2 - 17x_7 + 2x_8 \le 0$$
$$-3x_1 + 6x_2 + 12(x_9 - 8)^2 - 7x_{10} \le 0$$
$$-8x_1 + 2x_2 + 5x_9 - 2x_{10} \le 12$$
$$x_1 + x_2 + 4x_{11} - 21x_{12} \le 0$$
$$x_1^2 + 15x_{11} - 8x_{12} \le 28$$
$$4x_1 + 9x_2 + 5x_{13}^2 - 9x_{14} \le 87$$

6.3 TEST PROBLEMS

$$3x_1 + 4x_2 + 3(x_{13} - 6)^2 - 14x_{14} \leq 10$$

$$14x_1^2 + 35x_{15} - 79x_{16} \leq 92$$

$$15x_2^2 + 11x_{15} - 61x_{16} \leq 54$$

$$5x_1^2 + 2x_2 + 9x_{17}^4 - x_{18} \leq 68$$

$$x_1^2 - x_2 + 19x_{19} - 20x_{20} \leq 19$$

$$7x_1^2 + 5x_2^2 + x_{19}^2 - 30x_{20} \leq 0$$

Starting point: x_0 = (2, 3, 5, 5, 1, 2, 7, 3, 6, 10, 2, 2, 6, 15, 1, 2, 1, 2, 1, 3), $\varepsilon_1 = 10^{-4}$, $\varepsilon_2 = \varepsilon_3 = 10^{-2}$, $w_1 = w_{1max} = 0$, $w_2 = w_3 = 0.25$, $w_{2max} = w_{3max} = 8$, WF = 4, NSRCH = 41.

This function has a constrained local maximum at x^* = (2.04697, 2.20954, 8.74453, 5.06281, 0.956773, 1.43783, 1.33789, 9.97534, 8.17526, 8.45979, 2.31093, 1.35675, 6.10588, 14.1647, 0.998377, 0.495293, 1.49227, 2.00033, 2.64329, 2.02426) at which $f(x^*) = -130.490$, and β^* = (0.071786, 0, 0, 0.291890, 1.47694, 0.526358, 0, 1.45979, 0, 0.891876, 0.146436, 0, 0, 0.007967, 0.008488, 0, 0.134951).

Note: Some functions listed in [A8] differ slightly from corresponding ones in [W4]. The T14-T16 functions in this work are a fair compromise.

RESULTS FOR CONSTRAINED PROBLEMS

Table 6-3 lists the results obtained by applying the four modes of LPNLP to the constrained problems. In most cases, DFP performed better than DFP/Reset, and SS invariably performed better than SS/Reset. Comparing the SS modes to the DFP modes, we see that the DFP modes are generally superior; this is to be expected for those problems in which $N < 6$. For higher dimensioned problems (T15 with $N = 10$ and T16 with $N = 20$), SS shows an advantage over DFP/Reset, as expected; but contrary to expectations, DFP performs better than SS.

The test problems of this section are fairly well scaled, and there appears to be no advantage to the SS mode. When the number of variables and constraints of the problem increase, the structure of the Hessian is often changed after a multiplier and weight update in the early part of the maximization process, e.g., when an inequality constraint is changed from a locked-on to a free status or from a free to locked-on status. Thus, the self-scaling advantage may be lost here also.

Table 6-3

Results of Constrained Test Problems Using LPNLP Algorithm[1]

Prob.	N	expo	DFP ACC IT NF	DFP CONV IT NF	DFP/Reset ACC IT NF	DFP/Reset CONV IT NF	SS ACC IT NF	SS CONV IT NF	SS/Reset ACC IT NF	SS/Reset CONV IT NF
T7	3	-4	15 39	21 45	15 38	23 53	21 49	29 65	24 55	29 65
T8	4	-4	19 47	22 52	18 43	23 54	23 62	26 68	22 55	23 60
T9	7	-3	35 76	37 78	33 83	43 105	39 83	56 115	45 96	64 134
T10	2	-4	12 24	17 29	12 23	17 34	11 20	17 29	11 20	15 26
T11	4	-4	19 37	25 43	28 67	35 79	24 48	49 105	32 72	49 119
T12	3	-5	8 15	10 17	15 30	18 36	10 22	15 32	14 28	18 34
T13	5	-5	17 37	22 45	19 42	28 62	22 45	30 60	26 55	32 70
T14	7	-2	18 51	23 60	23 65	37 98	26 63	39 90	31 77	50 120
T15	10	-3	35 105	39 111	46 142	61 182	50 107	59 123	50 113	85 183
T16	20	-3	75 229	85 249	101 344	114 382	98 253	120 295	147 392	231 560

ACC (accuracy) -- conditions corresponding to a point at which $|f(x) - f(x^*)| < 10^{\text{expo}}$.
CONV (convergence) -- convergence based on $\|\nabla L_a\| < \varepsilon_1$ and $\|x^{k+1} - x^k\| < \varepsilon_3$. N -- number of variables. IT -- number of iterations. NF -- number of evaluations of the problem functions.
(1) Uses the unidirectional search of Section 5.4 and requires 1 gradient evaluation per iteration.

Test functions T10 through T16 are treated in a computational comparison study of some nonlinear programs [A8]. Three sequential unconstrained minimization methods, OPTKOV, POWCON, and SUMT, are compared on the basis of computer execution time and number of function evaluations. For test problems T10 through T16, the conclusions drawn from [A8] are: considering function evalutions, POWCON is the best of the three methods; considering execution time, SUMT is the best method; OPTKOV is relatively inefficient considering both number of function evaluations and runtime. The convergence test criteria for these three methods are not indicated in [A8], and therefore, the comparison with LPNLP is based on the number of function evaluations required to obtain a desired degree of accuracy in $f(x)$. The same starting points were used for each test problem. Function evaluations for OPTKOV, POWCON, and SUMT are obtained from a table in [A8], and are given in Table 6-4 with the results from Table 6-3 using the DFP mode without planned resets, as a basis for comparison.

6.4 TEST PROBLEMS

Table 6-4

Comparison of Several Methods

func.	N	ACC expo	Number of function calls[1]			
			LPNLP	SUMT	POWCON	OPTKOV
T10	2	-4	24	181	134	103
T11	4	-4	37	175	68[2]	172
T12	3	-5	15	>130	61	>133
T13	5	-5	37	126	90	187
T14	7	-2	51	>160[2]	123	334
T15	10	-3	105	532	272[3]	621
T16	20	-3	229	1120	270[3]	583[2]

(1) Corresponding to point where $|f(x) - f(x^*)| < 10^{expo}$.

(2) Accuracy achieved for $|f(x) - f(x^*)|$ was only 10^0.

(3) Accuracy achieved for $|f(x) - f(x^*)|$ was only 10^{-1}.

From the results given in Table 6-4, it is evident that the multiplier method presented in this text is superior to the penalty function methods studied in [A8].

6.4 A PROBLEM NOT SATISFYING SUFFICIENT CONDITIONS

The following problem is one that does not satisfy second-order sufficient conditions (see Example 4.2).

T17: (Rockafellar's problem [R3]).

Maximize

$$f = x_2 - x_1^4 - x_1 x_2$$

subject to $x_2 = 0$.

As is shown in Example 4.2, $x^* = (0, 0)$, $f(x^*) = 0$, and $\alpha^* = 1$. The augmented Lagrangian of equation (4-23) has a saddle point at $x = (0, 0)$, and maxima at $x = \pm((8w_1)^{-\frac{1}{2}}, -(32w_1^3)^{-\frac{1}{2}})$. As $w_1 \to \infty$, the maxima points approach $(0, 0)$. The Hessian of (4-23) is

$$\nabla^2 L_a \bigg|_{\alpha=\alpha^*} = \begin{bmatrix} -12x_1^2 & -1 \\ -1 & -2w_1 \end{bmatrix}$$

For $(-\nabla^2 L_a)$ to be positive definite, $x_1^2 > (24w_1)^{-1}$ but x_1 tends to zero only as w_1 tends to infinity, and therefore, the second-order sufficient conditions are not satisfied.

LPNLP was applied to solve this problem. Several solutions were obtained, each with a larger value of w_{1max}. A maximum iteration limit of 400 was set for each solution, and was initialized each time under the same conditions; $x_0 = (0.5, 0.5)$,

6.4 TEST PROBLEMS

Table 6-5

Results of Problem T17

w_{max}	x_1	x_2	α	$\|\nabla L_a\|$	f	IT	NF
10	$-1.5 \cdot 10^{-1}$	$1.5 \cdot 10^{-2}$	1.1536	$3.1 \cdot 10^{-1}$	$1.7 \cdot 10^{-2}$	$400^{(1)}$	1195
10^2	$5.6 \cdot 10^{-2}$	$-5.3 \cdot 10^{-4}$	1.0501	$1.7 \cdot 10^{-4}$	$-5.1 \cdot 10^{-4}$	$400^{(1)}$	1096
10^3	$-1.3 \cdot 10^{-2}$	$1.4 \cdot 10^{-5}$	0.9845	$5.1 \cdot 10^{-6}$	$1.5 \cdot 10^{-5}$	$400^{(1)}$	1307
10^4	$-7.7 \cdot 10^{-5}$	$-1.1 \cdot 10^{-14}$	1.0001	$2.2 \cdot 10^{-10}$	$-1.1 \cdot 10^{-14}$	$203^{(2)}$	590
10^5	$-8.0 \cdot 10^{-4}$	$-3.8 \cdot 10^{-9}$	1.0016	$5.8 \cdot 10^{-9}$	$-3.8 \cdot 10^{-9}$	$400^{(1)}$	1435
10^6	$-2.1 \cdot 10^{-4}$	$-2.0 \cdot 10^{-17}$	1.0002	$5.6 \cdot 10^{-11}$	$-2.1 \cdot 10^{-15}$	$103^{(2)}$	307
10^7	$-3.7 \cdot 10^{-5}$	$-2.6 \cdot 10^{-20}$	1.0000	$8.7 \cdot 10^{-13}$	$-1.8 \cdot 10^{-18}$	$136^{(2)}$	406
10^8	$-7.8 \cdot 10^{-6}$	$-1.0 \cdot 10^{-20}$	1.0000	$4.1 \cdot 10^{-12}$	$-1.4 \cdot 10^{-20}$	$131^{(2)}$	389

(1) Maximum iteration limit set at 400.
(2) Convergence criteria satisfied at this point.

$\varepsilon_1 = 10^{-9}$, $\varepsilon_2 = \varepsilon_3 = 10^{-4}$, $w_1 = 1$, $w_2 = w_3 = w_{2\max}$ = $w_{3\max}$ = 0, WF = 2, NSRCH = 5. The results are given in Table 6-5. It is seen that, with second-order sufficient conditions not satisfied at x^*, convergence is obtained by allowing the weights to approach ∞.

6.5 SUMMARY

Several results for the constrained test problems of this chapter were compared to solution results found in the literature using the penalty function techniques of sequential unconstrained optimization. Compared to the penalty techniques OPTKOV, POWCON, and SUMT, the multiplier technique LPNLP was shown to be a superior method, based on the number of function evaluations to obtain a desired degree of accuracy in f(x).

Tables 6-6 and 6-7 are based on test problem results of this chapter and are useful in summarizing the merits of the four operating modes of LPNLP. Table 6-6 contains solution statistics for average NF/IT, average number of function evaluations NF per iteration IT (function calls per

6.5 TEST PROBLEMS

Table 6-6

Function Calls per Iteration for Test Problems Solved by LPNLP (summarized from Tables 6-1 and 6-3)

Avg. NF/IT	DFP		DFP/Reset		SS		SS/Reset		
Problems	ACC	CONV	ACC	CONV	ACC	CONV	ACC	CONV	
unconstrained	2.46	2.45	2.42	2.45	2.31	2.31	--	--	
constrained	2.61	2.42	2.83	2.72	2.32	2.23	2.40	2.30	
	2.54		2.63		2.31		--		→Avg=2.49
All		2.44		2.59		2.27		--	→Avg=2.43

unidirectional search). Table 6-7 contains values of IT/N, ratios of the number of iterations (at convergence) to the number of variables N.

For the unconstrained problems, the SS mode of LPNLP yields both the smallest average value of NF/IT and that of IT/N. From these results, we conclude that the SS mode is generally the best solution mode for an unconstrained problem.

Of the two DFP modes, the NF/IT values of DFP/Reset are greater than those of DFP; but note that, for the unconstrained cases (Table 6-7a), the IT/N average for DFP/Reset is less than that for DFP. Of the fourteen values that combine to form the DFP average IT/N in Table 6-7a, those for T1(2), T2, and T6 appear to be excessive and may be the result of a slight and unpredictable tendency for search directions to approach linear dependence when periodic resets are not used with DFP.

For the constrained problems, the SS mode shows a slightly smaller NF/IT average, but this advantage is more than offset by the larger values of IT/N for SS in Table 6-7b. Of the two DFP modes, the DFP mode without automatic reset

6.5 TEST PROBLEMS

Table 6-7a

Iterations/Number-of-Variables at Convergence:
Summary for Unconstrained Problems

	DFP	DFP/Reset	SS
	IT/N	IT/N	IT/N
T1(2)	23.0	16.5	15.5
T1(6)	7.5	8.5	9.8
T1(10)	6.7	8.7	7.4
T1(16)	14.9	9.7	5.5
T1(30)	9.3	9.9	4.5
T1(50)	9.7	10.0	4.1
T1(100)	12.1	11.6	3.8
T2	22.5	17.3	21.5
T3	12.0	5.5	11.8
T4(10)	4.6	7.4	2.8
T4(30)	4.2	4.9	1.2
T4(50)	2.8	3.2	0.8
T5	4.7	5.7	5.3
T6	27.8	13.6	7.4
Average	11.56	9.46	7.24

Table 6-7b

Iterations/Number-of-Variables at Convergence: Summary for Constrained Problems

	DFP IT/N	DFP/Reset IT/N	SS IT/N	SS/Reset IT/N
T7	7.0	7.7	9.7	9.7
T8	5.5	5.8	6.5	5.8
T9	5.3	6.1	8.0	9.1
T10	8.5	8.5	8.5	7.5
T11	6.3	8.8	12.3	12.3
T12	3.3	6.0	5.0	6.0
T13	4.4	5.6	6.0	6.4
T14	3.3	5.3	5.6	7.1
T15	3.9	6.1	5.9	8.5
T16	4.3	5.7	6.0	11.6
Average	5.18	6.56	7.35	8.40

6.5 TEST PROBLEMS

performs consistently better on the constrained problems. Thus, despite some reservations prompted by the last statement in the preceding paragraph, the DFP mode is the recommended solution mode for constrained problems.

The values of IT/N in Table 6-7 are representative and can be used as a guide to selecting the search parameter IMAX for other problems. If the maximum number of allowed search iterations, IMAX, is set at 20·N, for example, the data of Table 6-7 would suggest that convergence will be obtained for most problems before the number of iterations equals IMAX. The value of IMAX = 20·N would not be adequate, however, for most problems like T17 (see Table 6-5).

PROBLEMS

6.1 A function similar to T1 with N = 2 is "Cube" [W1], $f = -100(x_1^3 - x_2)^2 - (1 - x_1)^2$ which has a steep cubic ridge along $x_2 = x_1^3$. Parallel the approach for test problem T1, and generalize Cube to N variables. Use the same parameter values as in T1 and use LPNLP to solve Cube for N = 6 and N = 30.

6.2 Let $r(x_1, x_2) = (x_1^2 + x_2^2)^{\frac{1}{2}}$ and $\theta(x_1, x_2) = \tan^{-1}(x_2/x_1)$.

i) $f_1 = 100(r-1)^2(r-2)^2 + 0.1(2-x_1)^2$ (Double)

ii) $f_2 = 1000(1 + \cos\theta - 0.1r)^2 + x_2^2(x_1-10)^4$

(Heart)

These functions are treated in [W1] where f_1 has 2 minima and 2 saddle points, and f_2 has 4 minima. Find starting points and initial conditions for LPNLP that will locate the maxima of $-f_1$ and $-f_2$.

6.3 Among the exponential functions considered by Box [B7] and Biggs [B6] is the Weibull function described by Shanno [S3]. The function to be maximized is

$$f = -\sum_{i=1}^{99} \{\exp[-(y_i - x_3)^{x_2}/x_1] - z_i\}^2$$

where $y_i = 25 + [50 \log_e(1/z_i)]^{2/3}$ and $z_i = (.01)i$, $i = 1, 2, \ldots, 99$. The initial starting point is $x^0 = (250, 0.3, 5)$.

6.4 Use LPNLP to solve the nonunique multiplier problem, Problem 2.10. Use several different starting points and explain your results.

6.5 From [Z4], maximize f,

$$f = -(x_1 - 2)^2 - (x_2 - 10)^2$$

subject to $x_1^2 + x_2^2 \le 73$ and $-x_1^2 - x_2^2 - 2x_1x_2 + x_1 + 11x_2 \le 36$. Use $x = (-4, 3)$ as the starting point for a computer solution.

6.6 From Lootsma [L2], the distance between a given sphere and a given cylinder is to be minimized. For the corresponding NLP, maximize f,

$$f = -\sum_{i=1}^{3} (x_i - x_{i+3})^2$$

subject to $x_1^2 + x_2^2 + x_3^2 \le 5$, $(x_4 - 3)^2 + x_5^2 \le 1$, and $4 \le x_6 \le 8$. Solve the problem using LPNLP. Use $\varepsilon_1 = 10^{-5}$, $\varepsilon_2 = \varepsilon_3 = 10^{-2}$, initial w_i's = 1, w_{imax}'s = 16, WF = 4, NSRCH = 13, and initial $x = (1, 1, 1, 3, 0, 5)$.

6.7 Of interest is the maximum of f,

$$f = (\sum_{i=1}^{8} x_i^2)(\sum_{i=1}^{8} x_i^4) - (\sum_{i=1}^{8} x_i^3)^2$$

subject to $0 \le x_i$'s ≤ 1. The global maximum of f is 1 and occurs at 80 distinct points. There are also 182 other local maxima. Use LPNLP with numerous different starting points to find points of global maximum. (This problem was posed by David Feder; explicit solutions are given by Wolfe [W3].)

6.8 Consider the following ill-conditioned NLP problem [C9]:

maximize $f = x_1 + x_2$

subject to $p(x) = 0$ where
$p(x) = \exp[(-x_1^2 - x_2^2 - 4)/4] - \exp[(-x_1^2 - x_2^2)/2]$.

a. Show that the constraint set is a circle with center at the origin and having a radius of 2. But note that $p(x) \to 0$ in the limit as $\|x\| \to \infty$.

b. Use LPNLP to solve the problem. Use several starting points, including the point (2, 0).

c. Modify the problem by replacing $p(x) = 0$ by $p(x) \leq 0$ and solve the modified problem using several starting points, including the point (0, 0).

6.9   Solve Problem 2.24 using LPNLP.

6.10   Solve Problem 4.13 using LPNLP.

6.11   Solve the following using LPNLP. Minimize f,

$$f = (x_1 - 1)(x_1 - 2)(x_1 - 3) + x_3$$

subject to $x_1^2 + x_2^2 \leq x_3^2$, $x_1^2 + x_2^2 + x_3^2 \geq 4$, $0 \leq x_3 \leq 5$, $x_1 \geq 0$, and $x_2 \geq 0$. This problem is considered in [F4].

6.12   Consider the following ill-conditioned system of three equations: $6x_1 + 13x_2 - 17x_3 = 1$, $13x_1 + 29x_2 - 38x_3 = 2$, and $-17x_1 - 38x_2 + 50x_3 = -3$. The solution is $x^* = (1, -3, -2)$. The system was constructed by E. Steifel [H5]. The ratio of the largest to the smallest eigenvalue is very large (λ_3/λ_1 = 1441). Use LPNLP to solve the above system starting at x^0 = (1, 0, 0).

6.13 Solve the following problem using LPNLP.

$$\text{Minimize } f = \sum_{k=1}^{10} kx_k^2$$

subject to

$$1.5x_1 + x_2 + x_3 + 0.5x_4 + 0.5x_5 = 5.5$$
$$2x_6 - 0.5x_7 - 0.5x_8 + x_9 - x_{10} = 2.0$$
$$x_1 + x_3 + x_5 + x_7 + x_9 = 10$$
$$x_2 + x_4 + x_6 + x_8 + x_{10} = 15$$

This problem is treated in [L6]. Hold the penalty weight w_1 constant for each run and study the behavior of the solution as w_1 increases (for example, w_1 = 1, 10, 100).

6.14 From Colville [C5], see also [A4], maximize

$$\sum_{j=1}^{5} e_j x_j + \sum_{j=1}^{5} \sum_{i=1}^{5} c_{ij} x_i x_j + \sum_{j=1}^{5} d_j x_j^3$$

subject to $x \geq 0$ and

$$\sum_{j=1}^{5} a_{ij} x_j \leq b_i, \quad i = 1, 2, \ldots, 10$$

where a_{ij}, c_{ij}, b_i, d_j, and e_j are given in Table
6.P14. Suggest values of LPNLP parameters to use,
and solve this problem using LPNLP with an initial
x = (0.125, 0.0625, 0.125, 0.125, 1.0). The solution is f = 32.34868 where x $\cong$ (0.3, 0.3335, 0.4, 0.4283, 0.2240).

Table 6.P14

	i\j	1	2	3	4	5	
e_j		15	27	36	18	12	
c_{ij}	1	-30	20	10	-32	10	
	2	20	-39	6	31	-32	
	3	10	6	-10	6	10	
	4	-32	31	6	-39	20	
	5	10	-32	10	20	-30	
d_j		-4	-8	-10	-6	-2	
							b_i
a_{ij}	1	16	-2	0	-1	0	40
	2	0	2	0	-0.4	-2	2
	3	3.5	0	-2	0	0	0.25
	4	0	2	0	4	1	4
	5	0	9	2	1	2.8	4
	6	-2	0	4	0	0	1
	7	1	1	1	1	1	40
	8	1	2	3	2	1	60
	9	-1	-2	-3	-4	-5	-5
	10	-1	-1	-1	-1	-1	-1

7
APPLICATION PROBLEMS

7.1 INTRODUCTION

In the first six chapters of this work, mathematical properties of a general NLP formulation have been emphasized. The theoretical framework for problem formulation, solution identification, and an implementation of this theory into a working algorithm have been presented and tested. We now focus on typical application problems, their transformation into explicit NLP's, and their solutions.

The NLP model can evolve in many forms from many unrelated areas. For example, Zangwill [Z2] treats the NLP formulation from problems that arise in production and inventory control, regression analysis, consumer behavior, chemical condensor design, rocket control, equation solving, cost-benefit analysis, financial analysis, and other aspects of economics, business, government, and engineering. Bracken and McCormick [B9] present several nonlinear problem formulations including models on weapons assignment, bid evaluation, alkylation process optimization, chemical equilibrium, structural optimization, launch vehicle design and costing, parameter estimation in curve fitting, optimal sample sizes in stratified sampling on several variates, and deterministic nonlinear programming equivalents for stochastic linear programming problems. Duffin et al. [D9] treat a wide class of problems with both fixed and adjustable parameters from the aspect of geometrical programming which is a special form of NLP particularly useful in engineering and the physical sciences. Steenbrink [S6] concentrates on problem formulations for optimizing transport networks. A variety of other economic, design, and decision making NLP models can be found in the literature [A9, F13, H2, S4, T2].

Optimal control is an active research area in modern technology, having applications in many areas, including aerospace, chemical, nuclear

7.1 APPLICATION PROBLEMS

reactor, and transportation technologies. Basic concepts of mathematical programming and the techniques for implementing solutions for optimal control problems are presented in [T1], along with examples and case studies in sub-areas of optimal control such as linear, nonlinear, continuous, and discrete-time systems as well as stochastic and distributed-parameter systems. Interesting approaches to digitally compensated control systems are given in [D4] and [P10].

For many years the practice for on-line control of power systems has been to approximate the nonlinear relations of such systems to obtain linear programming problems. The equations are then solved by well-established linear programming techniques. Application of nonlinear optimization for on-line control has the advantage of more accurate representation of the system, and the possibility of dealing with larger problems. Recent advances in this area are explored in [B10]. A variety of NLP models for power systems are available; examples include [A3, C1, D8, E1, S1].

Many approaches to the optimal design of physical structures [G1, S5] involve the use of nonlinear programming. Specific examples of NLP models of physical structures are given in [D7], [K1], and [N1].

Dennis [D6] investigates the application of mathematical programming in electrical networks. Trends in automated network design optimization as applied to computer-aided design of lumped-distributed and microwave networks are reviewed and discussed in [B1]. An NLP method for filter design is considered in [T3]. The problem of assigning tolerances to components in linear networks is cast in an NLP form in [P9] and [T4].

The above applications are typical, as are the three problem situations presented next. From each problem statement, an NLP model is formulated and then solved using the multiplier algorithm LPNLP. Solution results are interpreted in terms of the particular application.

7.2 A GEOMETRIC PROBLEM

The following problem is a geometric programming problem. It is an adaptation from [D9] and is presented in [Z2].

7.2 APPLICATION PROBLEMS

As vice president of a small but dynamic chemical company, you are to investigate containerization for product shipping. Not only should the containers simplify shipment, but also because the containers will be left with the customers, sales should be enhanced. Upon selecting a particular customer as a representative example of the problem, you find that each month 1000 cubic feet of a chemical must be shipped to this customer. The chemical is to be sent in rectangular containers of length x_1, width x_2, and height x_3. The thickness of all material used is negligible. The container seams are fused during the filling process to prevent moisture from entering. The sides and bottom must be made of scrap material for which there is no charge. However, only 10 square feet of this scrap can be guaranteed and processed per container each month from the materials department, but can be processed to any dimension. Material for the ends costs $2.00 per square foot, while material for the top costs $3.00 per square foot. There is also a shipping charge of 20 cents per container sent. You must determine how many and what size of containers are needed to ship the chemical at the lowest possible cost. Thus, the total cost (shipping plus the costs of the two ends and the top for $1000/x_1 x_2 x_3$ containers) must be minimized subject to the scrap limitation.

The total cost per container is

$$\text{cost} = (0.2 + 4x_2 x_3 + 3x_1 x_2) \text{ dollars} \qquad (7-1)$$

subject to the scrap limitation

$$2x_1 x_3 + x_1 x_2 \leq 10 \qquad (7-2)$$

Additional constraints must be added to ensure that the dimensions remain nonnegative.

$$x_1 \geq 0, \quad x_2 \geq 0, \quad x_3 \geq 0 \qquad (7-3)$$

Considering the total cost for all $(1000/x_1 x_2 x_3)$ containers, equation (7-1) becomes

$$\text{cost}_t = \frac{200}{x_1 x_2 x_3} + \frac{4000}{x_1} + \frac{3000}{x_3} \text{ (dollars)} \qquad (7-4)$$

Let $c_t(x)$ be the total cost in thousands of dollars. Because $c_t(x)$ is to be minimized, we define our objective function $f(x)$ as the negative

7.2 APPLICATION PROBLEMS

of $c_t(x)$. Then from (7-2), (7-3), and (7-4) we formulate the NLP as follows.

GEOMETRIC CONTAINER PROBLEM. Maximize

$$f = -c_t = -(\frac{0.2}{x_1 x_2 x_3} + \frac{4}{x_1} + \frac{3}{x_3}) \qquad (7-5)$$

in thousands of dollars, subject to

$$2x_1 x_3 + x_1 x_2 \leq 10 \qquad (7-6)$$

and

$$x_1 \geq 0, \quad x_2 \geq 0, \quad x_3 \geq 0 \qquad (7-7)$$

The optimum for this problem is at $x^* =$ (2.37976, 0.31623, 1.94294) at which $f(x^*) =$ -3.36168 and $\beta^* =$ (0.181762, 0, 0, 0).

For the solution, the algorithm was initialized as follows: $x_i =$ (1, 1, 1), $\varepsilon_1 = 10^{-5}$, $\varepsilon_2 = \varepsilon_3 = 10^{-2}$, $\beta_j = 0$, $j = 1, 2, 3, 4$, $w_1 = w_{1\max} = 0$, $w_2 = 1$, $w_3 = 50$, $w_{2\max} = 16$, $w_{3\max} = 100$, wf = 2, and nsrch = 7. The solution was obtained by all four search modes of LPNLP: DFP converged in 30 iterations and 106 function calls, DFP with reset in 26 and 87, selfscale in 41 and 143, and selfscale with reset in 32 and 117. The large initial weight for w_3 was to keep the values of x positive in the initial phases of the search (and to avoid constraint breakthrough in x), because f is unbounded if any one component of x approaches zero from negative values.

Interpreting the optimum of the NLP in terms of the given problem we find that 684 containers of length 2.380 feet, width 0.316 feet, and height 1.943 feet are required to ship 1000 cubic feet of chemical each month at a cost of $3362. The scrap limitation constraint is active (all 10 square feet per container are used) at the solution and has an associated positive multiplier of 0.181762. There is no shadow cost for the free scrap material. Therefore, if by some means, additional scrap material could be obtained at no cost, the incremental sensitivity indication is that an approximate savings of $182.00 from the total cost could be obtained for each additional square foot of scrap for each of the 684 containers. Reducing the above analysis to a single container basis, each container holds 1.46 cubic feet, and costs $4.92 for construction and shipping. For each additional

7.2 APPLICATION PROBLEMS

square foot of scrap per container, approximately 27 cents could be saved.

As seen in this problem, care must be taken in interpreting the results when initial scaling is performed on any of the problem functions. The units of f are thousands of dollars, and therefore, the units on the multiplier β are also in thousands of dollars, to keep consistency in the Lagrangian.

7.3 DETERMINISTIC NONLINEAR PROGRAMMING EQUIVALENT FOR A STOCHASTIC LINEAR PROGRAMMING PROBLEM

The problem presented here is equivalent to a stochastic linear programming problem of the chance-constrained type. In the chance-constrained problem, the constraint coefficients are normally distributed random variables. Charnes and Cooper [C3, C4] proposed the deterministic equivalent for the stochastic linear problem. The determination of least-cost optimal cattle feed using the associated NLP was modeled by van de Panne and Popp [V2], and is treated in [B9]. The presentation here of a least-cost optimal hog ration is treated in [L4] and parallels that in [B9].

The problem concerns the mixing of a number of raw materials in such a way that a hog ration is obtained that satisfies certain specified nutritive and other requirements with minimum cost for the input quantities of raw materials. If the nutritive contents and unit costs of raw materials and the requirements for the nutrients are known, the problem can be solved in a straightforward manner by deterministic linear programming methods. The problem that arises is that the nutritive content of raw materials varies randomly from batch to batch, so that the solution given by linear programming using expected (mean) values, for instance, does not always satisfy the requirements. That is, using the expected values from a normal distribution, there is only a 50% probability that the nutritive requirements of the ration will be satisfied.

Table 7-1 gives the data for a typical case. The percentage content of protein, calcium, and phosphorus are given for 13 constituents. The cost per ton (in hundreds of dollars) is given for each constituent. The requirements of protein, calcium, and phosphorus are given for starter pigs,

7.3 APPLICATION PROBLEMS

growers, and finishers. The problem is to determine the mix with minimum cost per ton that satisfies the given requirements. Since the method of solution is the same for all three rations, we will consider only the ration for starter pigs.

Let r_i, $i = 1, 2, \ldots, 13$, denote the cost per ton in hundreds of dollars of each constituent; and for each constituent, let s_i, u_i, and v_i, $i = 1, 2, \ldots, 13$, denote the percentage content for protein, calcium, and phosphorus, respectively. Let x_i, $i = 1, 2, \ldots, 13$, denote the fraction of the mixture that is composed of each of the constituents.

In order to compare the merits of both the linear and nonlinear models of this problem, solutions are obtained for both. First we formulate the linear problem using mean values.

CASE I (Linear Hog Ration). Again using the negative of the minimum cost function for the maximization problem, a deterministic linear programming model is as follows:

Table 7-1a
Data for least-cost hog ration problem

Fraction x_i i=1,2,...,13	Constituent	Cost/ton ($100) r_i	% Protein		% Calcium		% Phosphorus	
			mean $\bar{s}_i$	var. $\sigma_{s_i}^2$	mean $\bar{u}_i$	var. $\sigma_{u_i}^2$	mean $\bar{v}_i$	var. $\sigma_{v_i}^2$
x_1	barley	0.80	11.6	0.4844	0.05	0.0001	0.35	0.0010
x_2	wheat	1.10	13.7	0.3003	0.07	0	0.37	0.0009
x_3	corn	0.85	9.5	0.1444	0	0	0.10	0.0001
x_4	soy	3.45	48.5	0.0588	0.33	0	0.62	0.0005
x_5	mustard	2.00	31.9	4.9863	0	0	0	0
x_6	dried milk	2.10	51.1	0.0653	1.27	0.0040	1.03	0.0021
x_7	fish solubles	3.00	65.5	21.0222	1.27	0.1404	1.69	0.0825
x_8	d_i-cal.phos.	0.80	0	0	23.35	1.3631	18.21	0.2073
x_9	limestone	0.45	0	0	35.84	0.5138	0.01	0
x_{10}	molasses	0.72	0	0	0.81	0.0289	0.08	0.0004
x_{11}	dehy. alfalfa	1.80	21.8	0.2970	1.79	0.0097	0.31	0.0005
x_{12}	shrimp meal	3.00	46.9	9.2933	7.34	0.3893	1.59	0.0107
x_{13}	mono-sodium-phosphate	0.60	0	0	0	0	22.45	1.0206

7.3 APPLICATION PROBLEMS

Table 7-1b

Minimum content required in the feed

Type of hog	% Protein	% Calcium	% Phosphorus
starter	18	1.0	0.9
grower	16	0.8	0.7
finisher	14	0.7	0.6

Maximize

$$f = -\sum_{i=1}^{13} r_i x_i \qquad (7\text{-}8)$$

subject to

$$-\sum_{i=1}^{13} \bar{s}_i x_i \leq -18 \qquad (7\text{-}9a)$$

$$-\sum_{i=1}^{13} \bar{u}_i x_i \leq -1 \qquad (7\text{-}9b)$$

$$-\sum_{i=1}^{13} \bar{v}_i x_i \leq -0.9 \qquad (7\text{-}9c)$$

$$x_i \geq 0 \quad i = 1, 2, \ldots, 13 \qquad (7\text{-}9d)$$

$$\sum_{i=1}^{13} x_i = 1 \qquad (7\text{-}9e)$$

In the above deterministic model, we use the mean values of the s_i's, u_i's, and v_i's. The nonnegativity constraints (7-9d) are to ensure that the fractions composing the mixture remain greater than or equal to zero. That the sum of the parts equals the whole mixture is reflected in the equality constraint (7-9e).

The optimal solution is at $x^* = $ (0.785586, 0, 0, 0, 0, 0.173918, 0, 0, 0.020643, 0, 0, 0, 0.019853) at which $f(x^*) = -1.01490$, $\alpha^* = -0.417285$, and $\beta^* = $ (0.032743, 0.000911, 0.008138, 0, 0.231058, 0.120842, 1.43932, 0.538208, 0, 0.423127, 0.213247, 0, 0.301326, 0.664761, 1.02744, 0).

For the solution, the algorithm was initialized as follows: $x_i = 0.1$, $i = 1, 2, \ldots, 13$, $\varepsilon_1 = 10^{-4}$, $\varepsilon_2 = \varepsilon_3 = 10^{-2}$, $\alpha = \beta_j = 0$, $j = 1, 2, \ldots, 16$, $w_1 = w_2 = w_3 = 0.5$, $w_{1max} = w_{2max} = w_{3max} = 16$, wf = 2, and nsrch = 27. The solution was obtained by all four search modes of LPNLP. The algorithm converged using DFP in 51 iterations and 159 function calls, DFP with reset in 71 and 235, selfscale in 94 and 259, and selfscale with reset in 124 and 345.

Interpreting the optimum of the NLP in terms of the original problem, we find that a mixture of

7.3 APPLICATION PROBLEMS

x_1 = 0.7856 (barley)

x_6 = 0.1739 (dried milk)

x_9 = 0.0206 (limestone)

x_{13} = 0.0199 (mono-sodium phosphate)

will satisfy the minimum requirements of the ration for a cost of $101.49 per ton. But this ration will satisfy the requirement only 50% of the time, because the protein, calcium, and phosphorus percentage contents used were mean values. All three nutritive requirement constraints have associated positive Lagrange multipliers, and therefore, these requirements are all binding constraints at the solution.

CASE II (Nonlinear Chance-Constrained Formulation). The chance-constrained formulation of the linear problem treats the nutritive requirement constraints as follows [C3]:

$$P[\sum_{j=1}^{n} \rho_{ij} x_j \geq \xi_i] \geq \gamma_i \qquad (7\text{-}10)$$

where $P[\cdot]$ is the probability operator; some or all of the coefficients ρ_{ij} (i = 1, 2, ..., ℓ;

$j = 1, 2, \ldots, n$) are random variables with normal distributions; ξ_i are deterministic right-hand sides of constraints; and γ_i ($i = 1, 2, \ldots, \ell$) are prescribed probabilities with which the ℓ constraints must be satisfied.

If the ρ_{ij}'s in (7-10) are independent normally distributed random variables with means $\bar{\rho}_{ij}$'s and variances σ_{ij}^2's, inequality (7-10) can be shown to be equivalent to

$$\sum_{j=1}^{n} \bar{\rho}_{ij} x_j + \Psi(\gamma_i) [\sum_{j=1}^{n} \sigma_{ij}^2 x_j^2]^{\frac{1}{2}} \geq \xi_i \qquad (7-11)$$

where $\Psi(\gamma_i)$ is the percentage point or fractile corresponding to $(1 - \gamma_i)$, and is obtained from the inverse of the standardized normal left-tail cumulative function. For example, if the ith probability $\gamma_i = 0.95$, $\Psi(\gamma_i)$ is the 0.05 fractile. A value of $\gamma_i = 0.95$ corresponds to a value of $\Psi(\gamma_i) = -1.645$.

Inequality (7-11) is an appropriate relationship for the problem of this section. In a more general setting, however, the ρ_{ij}'s can be <u>dependent</u> multivariate normal variables, in which case (7-10) can be shown to be equivalent to

7.3 APPLICATION PROBLEMS

$$\sum_{j=1}^{n} \bar{\rho}_{ij} x_j + \Psi(\gamma_i)[x'Vx]^{\frac{1}{2}} \geq \xi_i \qquad (7\text{-}12)$$

where

$$V = \begin{bmatrix} \text{var}(\rho_{i1}) & \cdot & \cdot & \cdot & \text{cov}(\rho_{i1}\rho_{in}) \\ \cdot & \cdot & & & \cdot \\ \cdot & & \cdot & & \cdot \\ \cdot & & & \cdot & \cdot \\ \text{cov}(\rho_{in}\rho_{i1}) & \cdot & \cdot & \cdot & \text{var}(\rho_{in}) \end{bmatrix} \qquad (7\text{-}13)$$

Inequality (7-11) is a special case of (7-12), with the off-diagonal convariance terms in (7-13) equal to zero in (7-11).

Each of the stochastic constraints (7-10) must be considered in its nonlinear form (7-11). By using the means and variances listed in Table 7-1 in inequality (7-11), and assuming that the probability of satisfying each of the requirements is at least 0.95, we obtain the following nonlinear model of the least-cost hog ration problem.

Maximize

$$f = -\sum_{i=1}^{13} r_i x_i \qquad (7\text{-}14)$$

subject to

$$-\sum_{j=1}^{13} \bar{s}_j x_j - (-1.645)(\sum_{j=1}^{13} \sigma_{s_j}^2 x_j^2)^{\frac{1}{2}} \leq -18 \quad (7\text{-}15a)$$

$$-\sum_{j=1}^{13} \bar{u}_j x_j - (-1.645)(\sum_{j=1}^{13} \sigma_{u_j}^2 x_j^2)^{\frac{1}{2}} \leq -1 \quad (7\text{-}15b)$$

$$-\sum_{j=1}^{13} \bar{v}_j x_j - (-1.645)(\sum_{j=1}^{13} \sigma_{v_j}^2 x_j^2)^{\frac{1}{2}} \leq -0.9 \quad (7\text{-}15c)$$

$$x_j \geq 0, \quad i = 1, 2, \ldots, 13 \quad (7\text{-}15d)$$

$$\sum_{j=1}^{13} x_j = 1 \quad (7\text{-}15e)$$

A constrained local maximum for this problem is located at $x^* = (0.13205, 0, 0, 0, 0, 0.32627, 0, 0, 0.51668, 0, 0, 0, 0.025004)$ at which $f(x^*) = -1.03831$, $\alpha^* = -0.449929$, and $\beta^* = (0.032331, 0, 0.007135, 0, 0.204494, 0.092211, 1.42758, 0.518706, 0, 0.420314, 0.220149, 0, 0.269502, 0.643040, 1.02239, 0)$.

The algorithm was initialized the same as for the linear ration problem. The solution was obtained by all four search modes of LPNLP: convergence was obtained using DFP in 104 iterations and 265 function calls; DFP with reset in 154 and 394, selfscale in 140 and 415, and selfscale with reset in 214 and 599.

7.3 APPLICATION PROBLEMS

Interpreting the optimum of the NLP we find that a ration mixture of

$x_1 = 0.1320$ (barley)

$x_6 = 0.3263$ (dried milk)

$x_9 = 0.5167$ (limestone)

$x_{13} = 0.0250$ (mono-sodium phosphate)

will satisfy the minimum requirements for 95% of all batches mixed. The cost per ton is $103.83, which is an increase of approximately $2.34 over the linear program ration. But the $2.34 additional cost has bought a 95% probability that the ration requirements will be satisfied as compared to a 50% probability.

The mixture constituents (of the nonlinear model) are the same (as the linear model), although their mixture fractions have changed drastically. In the linear model, barley made up almost 80% of the mixture. In the nonlinear model, barley makes up only 13% of the mixture with dried milk and limestone making up the greatest percent of the mixture. Investigation of this mixture reveals that the calcium constraint is not binding at the solution. The actual amount of calcium in the mixture is approximately 18 times the amount required. If this is deemed to be too high a content for proper growth regulation in starter pigs, an additional constraint can be added to the problem giving an upper bound to the requirement. In fact, all requirements can be required to fall within certain upper and lower bounds. The requirements can also be forced to hold certain ratios to each other within the mixture by the addition of appropriate equality constraints.

Proper interpretation of the Lagrange multipliers gives an incremental indication of additional costs or savings that would result from requirement changes for the ration. The value of α^* has no sensitivity significance, in that the associated equality constraint is a mathematical law, stating that the percentage fractions of the constituents for the total ration must be 100% and cannot be changed. The associated multipliers with the $x_i \geq 0$ constraints indicate a savings for a small decrease from 0 in these constraints. But because the constituents of the ration cannot have a negative fraction, this sensitivity information is of no value. The first and third components of

7.3 APPLICATION PROBLEMS

β^*, those associated with the requirement constraints on protein and phosphorus, indicate a small savings could be obtained if these ration requirements could be reduced, i.e. lower the minimum ration requirements for protein and phosphorus. The first component is approximately 5 times larger than the second, indicating that the price per ton of the total ration is more sensitive to the requirement for protein than for phosphorus. But the magnitudes of both multipliers show relative insensitivity to both requirements for small changes, i.e. if the requirement for protein was reduced by one unit from 18 to 17, an approximate savings of $3.23 could be obtained on the price of the mixture (units of f are hundreds of dollars).

7.4 A CIRCUITS PROBLEM

The order of presentation in this section varies from that of the previous applications sections: first, an NLP with deceptively simple features is presented; next, computational results are given; and finally, an electrical circuit problem is reduced to the given NLP.

(A Circuits NLP). Maximize

$$f = -(r_1^2 + r_2^2) \qquad (7\text{-}16)$$

subject to

$$x_1 \geq 0.1, \quad x_2 \geq 0.1, \quad x_3 \geq 0, \quad x_4 \geq 0$$

and

$$(r_1^2 + r_2^2) - (r_3^2 + r_4^2) = 0 \qquad (7\text{-}17)$$

where

$$\begin{aligned}
r_1 &= 11 - x_1 x_4 - x_2 x_4 + x_3 x_4 \\
r_2 &= x_1 + 10 x_2 - x_3 + x_4 + x_2 x_4 (x_3 - x_1) \\
r_3 &= 11 - 4 x_1 x_4 - 4 x_2 x_4 + x_3 x_4 \\
r_4 &= 2 x_1 + 20 x_2 - 0.5 x_3 + 2 x_4 \\
 &\quad + 2 x_2 x_4 (x_3 - 4 x_1)
\end{aligned} \qquad (7\text{-}18)$$

The constrained maximum for this problem is as follows: $x^* = (1.91663, 0.1, 0, 1.97181)$, $f(x^*) = -69.6755$, $\alpha^* = 0.197551$, and $\beta^* = (0, 10.7852, 11.4656, 0)$.

Several different initial values of x and different values of nsrch were used in solving the problem. Other search parameters were initialized

7.4 APPLICATION PROBLEMS

as follows: $\varepsilon_1 = 10^{-4}$, $\varepsilon_2 = \varepsilon_3 = 10^{-2}$, $\alpha = \beta_j = 0$, $j = 1, 2, 3, 4$, $w_1 = w_2 = 0.5$, $w_3 = 4$, $w_{1max} = w_{2max} = 50$, $w_{3max} = 100$, and $wf = 4$.

Computational results are given in Table 7-2. Note that the DFP mode gives both the best result (147, 327) and the worst result (993, 2789). The latter result suggests a tendency for linear dependencies of search directions to occur when planned resets to the gradient direction are not included. The DFP/Reset mode is the most consistent for this problem and has the best average of 203.4 iterations for convergence. However, when this average of 203.4 is divided by 4 (the number of variables), the result 50.85 is significantly larger than corresponding ratios for the test problems of Chapter 6. Computational difficulties are clearly evident for this problem.

The above NLP stems from the following electrical circuits problem. Consider the circuit shown in Figure 7-1 (Problem 6.31 of [P4]). The ratio $|V_2(j\omega)/V_1(j\omega)|$ is of interest at two given frequencies $\omega = \omega_1 = 10^6$, and $\omega = \omega_2 = 2 \cdot 10^6$. $|V_2(j\omega_1)/V_1(j\omega_1)|$ is to be maximized subject to

Table 7-2

Convergence Results For The Circuits NLP

Conditions	DFP IT[1]	DFP NF	DFP/Reset IT	DFP/Reset NF	SS IT	SS NF	SS/Reset IT	SS/Reset NF
initial x = (1, 1, 1, 1) NSRCH = 9	993	2789	191	622	351	1061	484	1633
x = (1, 1, 1, 1) NSRCH = 21	555	1439	216	592	351	1021	317	1073
x = (3, 1, 1, 3) NSRCH = 9	197	487	155	462	153	373	183	563
x = (3, 1, 1, 3) NSRCH = 21	147	327	220	606	209	463	275	775
x = (0, 0, 0, 0) NSRCH = 21	208	555	235	708	459	1369	285	862

(1) IT = Number of unidirectional searches, NF = number of evaluations of the model.

7.4 APPLICATION PROBLEMS

the constraint that $|V_2(j\omega_1)/V_1(j\omega_1)| = |V_2(j\omega_2)/V_1(j\omega_2)|$. Here $j = (-1)^{\frac{1}{2}}$. Components $R_1 = 10K\Omega$ and $R_2 = 1K\Omega$ are fixed; C_1, C_2, L_1, and L_2 are to be selected to give the constrained optimum. Additional constraints are $L_1 \geq 0$, $L_2 \geq 0$, $C_1 \geq 10$ picofarad, and $C_2 \geq 100$ picofarad.

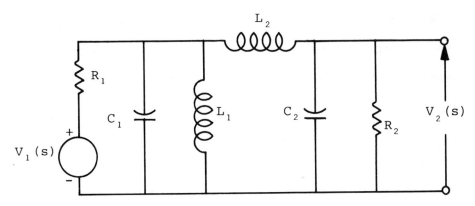

Figure 7-1 Circuit for NLP formulation.

For reasons that will be made clear later, this maximization problem may be put into the NLP format by considering the following equivalent minimization problem:

$$\text{minimize} \quad \left|\frac{V_1}{V_2}\right|^2_{\omega=\omega_1} \tag{7-19}$$

subject to

$$\left|\frac{V_1}{V_2}\right|^2_{\omega=\omega_1} - \left|\frac{V_1}{V_2}\right|^2_{\omega=\omega_2} = 0$$

$$L_1 \geq 0, \quad L_2 \geq 0, \tag{7-20}$$

$$C_1 \geq 10^{-11}, \quad \text{and} \quad C_2 \geq 10^{-10}$$

Relations (7-19) and (7-20) cannot be solved directly in this form. The elements L_1, L_2, C_1, and C_2 must be related to V_1, V_2, and ω through appropriate circuit relations. Proper reduction of this circuit by the use of Laplace transform theory yields the following relation:

$$\frac{V_1(s)}{V_2(s)} = H_1 H_2 - (R_1/sL_2) \tag{7-21}$$

where

$$H_1 \triangleq 1 + sR_1C_1 + \frac{R_1}{sL_1} + \frac{R_1}{sL_2}, \quad \text{and}$$

$$H_2 \triangleq 1 + s^2 L_2 C_2 + \frac{sL_2}{R_2}$$

We define

$$a \triangleq R_1/R_2 = 10 \qquad b \triangleq \omega_2/\omega_1 = 2 \tag{7-22a}$$

$$x_1 \triangleq \omega_1 R_1 C_1 \qquad\qquad x_2 \triangleq \omega_1 R_2 C_2 \tag{7-22b}$$

7.4 APPLICATION PROBLEMS

and

$$x_3 \triangleq R_1/\omega_1 L_1 \qquad x_4 \triangleq \omega_1 L_2/R_2 \qquad (7\text{-}22c)$$

so that

$$\begin{aligned} bx_1 &= \omega_2 R_1 C_1 & bx_2 &= \omega_2 R_2 C_2 \\ x_3/b &= R_1/\omega_2 L_1 & bx_4 &= \omega_2 L_2/R_2 \end{aligned} \qquad (7\text{-}23)$$

Let $s = j\omega$. If we substitute the identities of (7-22) into (7-21) and collect terms, we have

$$\frac{V_1(j\omega_1)}{V_2(j\omega_1)} = [a + 1 - x_1 x_4 - x_2 x_4 + x_3 x_4]$$
$$+ j[x_1 + ax_2 + x_4 - x_1 x_2 x_4 \qquad (7\text{-}24)$$
$$+ x_2 x_3 x_4 - x_3]$$

We see from (7-24) that the ratio $V_1(j\omega_1)/V_2(j\omega_1)$ is a complex number, having both real (Re) and imaginery (Im) components. Similarly, a second ratio is obtained when (7-21) is evaluated at $\omega = \omega_2$ using the identities of (7-23):

$$\frac{V_1(j\omega_2)}{V_2(j\omega_2)} = \text{Re}(\omega_2) + j\text{Im}(\omega_2) \qquad (7\text{-}25)$$

If we use the fact that

$$\left|\frac{V_1}{V_2}\right| = (Re^2 + Im^2)^{1/2} \tag{7-26}$$

then by substituting the Re and Im of (7-24) into (7-26) and squaring we have, with $a = 10$,

$$\left|\frac{V_1}{V_2}\right|^2_{\omega=\omega_1} = r_1^2 + r_2^2 \tag{7-27}$$

where

$$r_1 = 11 - x_1 x_4 - x_2 x_4 + x_3 x_4, \text{ and}$$

$$r_2 = x_1 + 10 x_2 - x_3 + x_4 - x_1 x_2 x_4$$

$$+ x_2 x_3 x_4$$

Similarly, upon solving (7-25) in terms of x and substituting into (7-26) and then squaring we have

$$\left|\frac{V_1}{V_2}\right|^2_{\omega=\omega_2} = r_3^2 + r_4^2 \tag{7-28}$$

where

$$r_3 = 11 - 4 x_1 x_4 - 4 x_2 x_4 + x_3 x_4$$

and

7.4 APPLICATION PROBLEMS

$$r_4 = 2x_1 + 20x_2 - 0.5x_3 + 2x_4$$
$$- 8x_1x_2x_4 + 2x_2x_3x_4$$

From the definition of x_1 in (7-22) and from the relations in (7-20), we know that

$$C_1 = \frac{x_1}{\omega_1 R_1} = \frac{x_1}{10^{10}} \geq 10^{-11} \qquad (7-29)$$

from which

$$x_1 \geq .1 \qquad (7-30)$$

Similarly we find that

$$x_2 \geq .1, \quad x_3 \geq 0, \quad x_4 \geq 0 \qquad (7-31)$$

From the above development, the NLP can be stated in familiar form. We collect equations (7-27) and (7-28), and relations (7-30) and (7-31), and upon comparing them with (7-19) and (7-20), we have the NLP of (7-16), (7-17), and (7-18).

Interpreting the results of the NLP in terms of the given problem, we find that the voltage ratio $|V_2(j\omega_1)/V_1(j\omega_1)|$ has the same maximum value of $(69.6755)^{-\frac{1}{2}} = 0.11980$ at the two frequencies $\omega_1 = 10^6$ and $\omega_2 = 2 \cdot 10^6$. This value is obtained by the selection of $C_1 = 192$ picofarads, $C_2 = 100$ picofarads, $L_1 = \infty$, and $L_2 = 1.97$ millihenries. The value $L_1 = \infty$ indicates that the inductor L_1 should not be included in the circuit.

The equality constraint multiplier α^* has no sensitivity meaning if the voltage ratios must be exactly the same at the two given frequencies. However, we see from the small value of α^* ($\alpha^* \simeq 0.2$), that the circuit is relatively insensitive to small differences in the voltage ratios at the local maximum. The multiplier associated with L_1 has no sensitivity meaning in that an incremental change in the value of ∞ has no significance. The multiplier associated with the inequality constraint on C_2 has some slight significance. The incremental sensitivity indication is that if the right side of $x_2 \geq .1$ could be decreased by a small amount, say $x_2 \geq 0$, a reduction of $(0.1)(10.7851)$ would occur in the value of $|V_1(j\omega_1)/V_2(j\omega_1)|^2$, and thus the new value of $|V_2(j\omega_1)/V_1(j\omega_1)|$ would be $(69.6755 - 1.0785)^{-\frac{1}{2}} = 0.12074$. This is less than a 1% improvement in the voltage ratio.

7.5 SUMMARY

Three diverse application problems were presented in this chapter. The first was a geometric containerization problem, in which similar rectangular containers were designed to ship a given amount of chemical, for minimum total cost. The second problem was a direct application to the livestock industry. Two minimum-cost feed rations for hogs were obtained: one from a linear model and one from a nonlinear model. The merits of the results from both models were evaluated.

The third was a circuit design problem in which design-center values of certain components were to be selected to maximize the output voltage to the same value for two given frequencies.

The three problems were presented to illustrate how NLP's may be formulated from real-world problem situations, and how the results may be interpreted and implemented. The importance of the Lagrange multipliers was seen through the interpretation of the sensitivity information they contain.

PROBLEMS

7.1 Consider the problem of maximizing f,

$$f = 2y_1 + 0.1y_2 + 10y_3$$

subject to $y_1, y_2, y_3 \geq 0$ and

$$y_1^2 + y_1 - 0.2y_2 - 10y_3 \leq -1$$

$$200y_1 - 10y_2 + 2000y_3 \leq 0$$

$$10y_1 + y_2 + 100y_3 \leq 40$$

    a. Suggest multiplicative factors for the last two inequality constraints, with the objective of making the coefficients of the y_i's the same order of magnitude as the corresponding coefficients in f.

    b. Suggest scale changes for the variables: let $y_i = \gamma_i x_i$ and select γ_i's so that the coefficients of x_i's in f are of the same order of magnitude.

    c. Solve the scaled problem and determine the optimal y_i's. How much will f^* change if the right-hand sides of the active inequality constraints are increased by 10%?

7.2 Solve the NLP of Problem 1.1 using $(a_1, a_2, a_3) = (5000, 8000, 7000)$ and $(b_1, b_2, b_3) = (0.1, 0.07, 0.08)$.

7.3 Solve the least-squares problem, Problem 1.3, using the following values of $z(r_1, r_2)$:

	r_2				
r_1 \	1	2	3	4	5
1	12.5	22.0	25.5	24.5	16.0
2	15.0	26.5	30.5	30.0	23.5
3	14.5	26.0	32.0	31.0	26.5
4	9.0	22.5	28.5	30.0	25.0
5	0.5	14.5	22.0	23.5	20.5

7.4 Consider the investment problem, Problem 1.4, with the following values:

i	1	2	3	4	5
$\bar{r}_i$	1.05	1.10	1.15	1.30	2.0
$\sigma^2_{r_i}$	0.0	0.001	0.007	0.1	1.0

a. Use LPNLP to solve the problem as stated.

b. Use LPNLP to solve the problem when only 90% confidence of 105% or better value is required (here $\Psi(\gamma) = -1.28155$).

7.5 Consider the following problem from [T2]. Convert the following stochastic problem into an equivalent deterministic model and solve.

Maximize $f = x_1 + 2x_2 + 5x_3$

subject to

$P\{a_1 x_1 + 3x_2 + a_3 x_3 \leq 10\} \geq .9$

$P\{7x_1 + 5x_2 + x_3 \leq b_2\} \geq .1$

and $x_1, x_2, x_3 \geq 0$. Assume a_1 and a_3 are independent normally distributed random variables with means $\bar{a}_1 = 2$, $\bar{a}_3 = 5$, and variances $\sigma_1^2 = 9$, $\sigma_3^2 = 16$. Assume further that b_2 is normally distributed with mean $\bar{b}_2 = 15$ and variance $\sigma_2^2 = 25$.
(Hint: see Case II of Section 7.3.)

7.6 In planning for production of two products, company Z estimates profit to equal f (in units of 10^5 dollars),

$$f = 3(1 - e^{-1.2x_1} - 1.2x_1 e^{-1.2x_1})$$

$$+ 4(1 - e^{-1.5x_2} - 1.5x_2 e^{-1.5x_2})$$

$$+ (1 - e^{-x_1 x_2})\gamma - x_1 - x_2$$

where x_1 is the amount of money spent to produce and advertise product 1, and x_2 is that spent on product 2. The units of x_1 and x_2 are 10^5 dollars. Constraints are $(x_1, x_2) \geq 0$, and $x_1 + x_2 \leq 3$. Use LPNLP to solve for the optimum under the following two conditions:

    a. $\gamma = 1$

    b. $\gamma = -1$ (several local maxima exist)

7.7 Use LPNLP to solve the conveyor problem, Problem 1.6, under the following conditions: $h = 10$ meters, $\ell = 20$ meters, and $g = 9.8$ m/sec^2.

7.8 Generalize Problem 1.6 to include viscous friction: motion along a straight segment of the slide is characterized by $\ddot{s} + \gamma \dot{s} = (\sin \theta)/g$. Use $h = 10$, $\ell = 20$, $g = 9.8$, and solve the problem

APPLICATION PROBLEMS 293

using several values of γ, $\gamma = 0.1$, 0.05, and 0.01. Over any straight section, velocity $\dot{s}(t)$ equals $\dot{s}(t_0)e^{-\gamma(t-t_0)} + (1/\gamma g)(\sin\theta)[1 - e^{-\gamma(t-t_0)}]$.

7.9 As in [M9], maximize system reliability R_s,

$$R_s = \prod_k [1 - (1 - r_k)^{x_k}]$$

subject to

$$\sum_k a_k x_k \exp[b_k/(1 - r_k)] \le C \text{ (cost)}$$

$$0 \le r_k \le 1, \quad k = 1, 2, \ldots, m$$

where a_k's and b_k's are constants, x_k is the number of redundant units in stage k, and r_k is the reliability of each unit in stage k. Both x_k's and r_k's are to be selected to obtain the optimum. (x_k's are to be rounded to the nearest integer after the continuous NLP solution is obtained.) Use $C = 300$, $m = 4$, and the following data:

k	1	2	3	4
a_k	1.00	3.50	2.00	5.00
b_k	0.30	0.55	0.40	0.65

7.10 A transversal digital filter is characterized by $G(z) = x_1 z^{-1} + x_2 z^{-2} + \cdots + x_{n-1} z^{-(n-1)}$ where: x_i's are real values to be selected; $z = e^{j\omega T} \triangleq e^{j\gamma}$; $j = \sqrt{-1}$, and γ assumes values from 0 to π. For γ in the interval $(0, \pi/4)$, we are given that $|G(e^{j\gamma})| \simeq 1$ is desired; and for γ in the interval $(\pi/4, \pi)$, $|G(e^{j\gamma})| \simeq 0$.

Consider the following constraints:

$$|G(e^{j\gamma})|^2 \leq (1 + x_n)^2,$$

$$(4\gamma/\pi) = 0, 0.1, \ldots, 0.9$$

$$|G(e^{j\gamma})|^2 \geq (1 - x_n)^2,$$

$$(4\gamma/\pi) = 0, 0.1, \ldots, 0.9$$

and

$$|G(e^{j\gamma})|^2 \leq x_n^2,$$

$$(4\gamma/\pi) = 1.2, 1.4, \ldots, 4.0$$

A minimum of x_n is desired. Give the corresponding NLP and solve it for $n = 6$ and $n = 10$.

7.11 Generalize Problem 1.1 to the case where company Z must produce 10,000 units of product 1 and 15,000 units of product 2 during a given time period. Give a detailed description of your generalized problem.

7.12 Reconsider Problem 1.5. Specify realistic $g_1(x_1,k)$ and $g_2(x_2,k)$ functions for $k = 1, 2, \ldots, 10$. Let $r_1 = 0.08$ and $r_2 = 0.07$, and solve the problem using your g_1's and g_2's.

APPENDIX A

A.1 GENERAL NOTATION

Subroutine LPNLP is a FØRTRAN coded routine that solves the general nonlinear programming problem (NLP) which is of the form:

$$\max f(x) \tag{A-1}$$

subject to the constraints

$$p_i(x) = a_i, \quad i = 1, 2, \ldots, m_1 < n \tag{A-2}$$

$$q_j(x) \leq b_j, \quad j = 1, 2, \ldots, m_2 \tag{A-3}$$

and

$$c_k \leq x_k \leq d_k, \quad k = 1, 2, \ldots, n \tag{A-4}$$

The notation used in the FØRTRAN representation of the above problem is given in Table A-1. The kth variable x_k is represented by X(K); the objective function f is represented by F; FE(I) is assigned the ith equality-constraint function $p_i(x)$ that is

Table A-1

FØRTRAN Representation of Terms Contained in (A-1) to (A-4)

Mathematical Representation	FØRTRAN Representation
n	N
x_k, $k = 1, 2, \ldots, n$	X(K), K = 1, 2, ..., N
$f(x)$	F
$p_i(x)$, $i = 1, 2, \ldots, m_1$	FE(I), I = 1, 2, ..., NE
(e.g., $p_2(x) = 2x_1 + x_2$)	(FE(2) = 2.*X(1) + X(2))
a_i	A(I)
$q_j(x)$, $j = 1, 2, \ldots, m_2$	FI(J), J = 1, 2, ..., NI
(e.g., $q_3(x) = x_1^2 + x_2$)	(FI(3) = X(1)**2 + X(2))
b_j	B(J)
c_k	C(K)
d_k	D(K)
$\partial f / \partial x_k$	GF(K)
$\partial p_i / \partial x_k$	GE(N*I − N + K)
$\partial q_j / \partial x_k$	GI(N*J − N + K)

A.1 INSTRUCTIONS FOR USING LPNLP

constrained to equal the constant A(I); FI(J) is assigned the jth inequality-constraint function $q_j(x)$ that is constrained to be less than or equal to B(J); and each X(K) can be assigned an explicit lower bound C(K) and an explicit upper bound D(K) -- NBV is the number of variables that are assigned an upper bound or a lower bound or both.

The FE(I)'s, FI(J)'s, and F are given functions of class C^1, and may be linear or nonlinear. Any of the index limits NE, NI, or NBV may be equal to zero. If all index limits are equal to zero, the problem is unconstrained.

The general routine needs no modification from problem to problem, but a short calling program, two subroutines, and a data deck do require modification by the user from problem to problem as they supply both the analytical representation of a particular problem and the information dictating operating modes and initial conditions of the general routine for that problem. An explanation of the user supplied routines and data will now be given.

A.2   USER SUPPLIED ROUTINES

A brief description and form listing of each subroutine is given in this section.

MAIN CALLING PROGRAM. The user supplied main program properly dimensions all vectors used by the general routine LPNLP, and makes the appropriate call to LPNLP, thus passing control to the general routine. The calling program is:

```
      DIMENSIØN DLX(N),DLG(N),GAL(N),GF(N),GØ(N),
     &IX(N),R(N),X(N),XT(N),H(N*N)
      DIMENSIØN A(NE),FE(NE),FES(NE),HE(NE),GE(N*NE)
      DIMENSIØN B(NI),FI(NI),FIS(NI),HI(NI),GI(N*NI)
      DIMENSIØN C(N),D(N),HVL(N),HVU(N)
      CALL LPNLP(A,B,C,D,DLX,DLG,FE,FES,FI,FIS,
     &GAL,GE,GF,GI,GØ,H,HE,HI,HVL,HVU,IX,R,X,XT)
      STØP
      END
```

The arguments in the dimension statements are symbolic; specific integer <u>values greater than or equal to these values</u> must be used in the dimension statements when executing a program. N is the number of x_i variables, NE is the number of equality constraints, and NI is the number of inequality constraints. If there are no equality constraints (i.e., if NE = 0), the vectors A, FE,

A.2 INSTRUCTIONS FOR USING LPNLP

FES, HE, and GE need not be dimensioned. If there are no inequality constraints (i.e., NI = 0), the vectors B, FI, FIS, HI, and GI need not be dimensioned. If none of the X(K) variables are bounded in any way (i.e., if NBV = 0) such that X(K) can take on any value in the range $(-\infty, \infty)$, the vectors C, D, HVL, and HVU need not be dimensioned. Since the argument list of LPNLP is long, the user is cautioned to check that precisely 24 arguments are listed. All arguments must be listed regardless of the number of vectors dimensioned above. In the remaining user supplied subroutines, all vectors are given the dimension (1), as allowed by FØRTRAN.

SUBROUTINE FOR SUPPLYING PROBLEM FUNCTIONS. The user supplied subroutine FXNS lists the objective function F to be maximized, and the left-hand side FE of all equality constraints, and the left-hand side FI of all inequality constraints. The right-hand side of the equality and inequality constraints are the constant vectors A and B of the equations and are supplied through the data deck. The bounded variables and their upper and lower bounds are also supplied through the data deck. By supplying the constant terms of the equations through the data deck, a sensitivity analysis can be conducted and the problem re-run without recoding the equations listed in this subroutine. The subroutine FXNS is of the following form:

```
SUBRØUTINE FXNS(X,F,FE,FI)
DIMENSIØN X(1),FE(1),FI(1)
F        =  ..........  cost function
FE(1)    =
  •
  •      =  ..........  NE equality
  •                     constraints; if
                        NE = 0, no listing
FE(NE)   =
FI(1)    =
  •
  •      =  ..........  NI inequality
  •                     constraints; if
                        NI = 0, no listing
FI(NI)   =
RETURN
END
```

SUBROUTINE FOR SUPPLYING THE NONCONSTANT GRADIENT COMPONENTS[†] OF THE PROBLEM FUNCTIONS.

This user supplied gradient subroutine gives the nonconstant gradient components GF, GE, and GI of

---

[†] Nonconstant gradient components are those that depend upon x in any way, as opposed to those which are strictly constants. For example, $q_3 = x_1^2 + x_1 + 5x_2$ has the nonconstant gradient component $\partial q_3/\partial x_1 = 2x_1 + 1$, and the constant gradient component $\partial q_3/\partial x_2 = 5$.

A.2 INSTRUCTIONS FOR USING LPNLP

the objective function F, and of all equality and inequality constraints, FE and FI. The constant gradient components are supplied through the data deck. This routine is called after every unidirectional search when a new X point is found. If there are no nonconstant gradient components (as in the case of a linear programming problem), the SUBRØUTINE, DIMENSIØN, RETURN, and END statements must still be supplied by the user. This routine is of the form:

```
      SUBRØUTINE GRAD(X,GF,GE,GI)
      DIMENSIØN X(1),GF(1),GE(1),GI(1)
      GF(1)    =              N components of the
        .                     gradient of F. If a
        .                     given (total) GF com-
        .                     ponent is a constant,
        .      = ..........   it need not be list-
        .                     ed here. If there are
        .                     no nonconstant GF com-
        .                     ponents, no listing
      GF(N)    =              is needed.
      GE(1)    =
        .                     N components of the
        .                     gradient of FE(1),
        .                     then N such components
        .      = ..........   for FE(2), ..., then
        .                     N such components for
        .                     FE(NE). If a given
        .                     (total) GE component
        .                     is a constant, it need
      GE(N*NE) =              not be listed here.
      CØNTINUE                If NE = 0, no list-
                              ing is needed.
```

```
          GI(1)     =                    N components of the
            .                            gradient of FI(1),
            .                            then N such compo-
            .                            nents for FI(2), ...,
            .                            then N such compo-
            .       = ..........         nents for FI(NI). If
            .                            NI = 0, no listing is
            .                            needed. If a given
            .                            (total) GI component
            .                            is a constant, it
            .                            need not be listed
          GI(N*NI)  =                    here.
          RETURN
          END
```

All gradient components equal to zero need not be specified. The LPNLP routine automatically zeroes all gradient components before initializing to the constant components through the data deck and before the first call to subroutine GRAD.

A.3 INFORMATION AND FORMAT OF THE DATA CARDS

The data cards communicate to LPNLP the initial conditions of certain variables, flags and parameters, convergence and update criteria, operating modes, stopping conditions, and print intervals associated with the output.

The correct order and format of the data cards, with parameter definitions and typical values where applicable, are as follows:

A.3 INSTRUCTIONS FOR USING LPNLP

DATA CARD 1:

```
      READ 1170,(ID(I),I=1,18)
1170  FØRMAT (18A4)
```

This is a user identification card to be printed at the beginning of the output. It is a problem title card containing such information as: problem number, book, author, or any other description such as linear, nonlinear, 6th-order, constrained, unconstrained, etc.

DATA CARD 2:

```
      READ 1000,N,NE,NI,NBV,IMAX,NPRINT,NUP,NSRCH,
     &ISS,IRESET
1000  FØRMAT (10I5)
```

    N = number of variables.

   NE = number of equality constraints.

   NI = number of inequality constraints.

  NBV = number of variables that are bounded in any way (number of variables for which finite c_k's or d_k's or both are assigned).

IMAX = maximum iteration limit on the number of unidirectional searches; LPNLP stops when IMAX searches have been made.

NPRINT = print interval: initial values and the final search point information are printed automatically by LPNLP. If this is all that is desired, set NPRINT = 0. If search point information is desired every J searches, set NPRINT = J, where $J \in \{1, 2, \ldots, \text{IMAX}\}$.

NUP = 0 causes the update information to be printed when the multipliers and weights are updated when solving a constrained problem.
  = 1 suppresses update information.

NSRCH = the maximum number of unidirectional searches in each cycle. Multipliers and weights are updated at the end of a cycle. Typically NSRCH = (2N+1).

ISS = 0 gives DFP search-direction mode.
  = 1 gives DFP/SS search-direction mode. See the test problem results of Chapter 6.

IRESET = 0 produces no reset to gradient direction at the end of a cycle.
  = 1 produces a reset to the gradient direction at the end of a cycle, provided at least N unidirectional searches have been made since the last reset. See the test problem results of Chapter 6 for suggested settings for ISS and IRESET.

DATA CARD 3:

```
      READ 1000,NA, NB,IGF,IGE,IGI,IHE,IHI,IHVL,IHVU
1000  FØRMAT (10I5)
```

NA = number of equality constraints which have a nonzero right-hand side (number of $A(I) \neq 0$).
  = 0 if NE = 0 or if A(I) = 0 for all I.

A.3 INSTRUCTIONS FOR USING LPNLP

NB = number of inequality constraints, those associated with FI's, which have a nonzero right-hand side (number of $B(J) \neq 0$).
 = 0 if NI = 0 or if $B(J) = 0$ for all J.

IGF = number of nonzero constant gradient components GF, of the cost function F, that are to be read from the data deck. IGF $\in [0,N]$.
 = 0 if there are no nonzero constant gradient components GF to be read from the data deck.

IGE = number of nonzero constant gradient components GE, of the equality constraints FE, that are to be read from the data deck. IGE $\in [0,N*NE]$.
 = 0 if NE = 0 or if there are no nonzero constant gradient components GE to be read from the data deck.

IGI = number of nonzero constant gradient components GI, of the inequality constraints FI, that are to be read from the data deck. IGI $\in [0,N*NI]$.
 = 0 if NI = 0 or if there are no nonzero constant gradient components GI to be read from the data deck.

IHE = number of equality constraint multipliers HE that are to be initialized to some nonzero constant (number of $HE(I) \neq 0$ initially). IHE $\in [0,NE]$.
 = 0 if NE = 0 or if all equality constraint multipliers are to be zero initially (this is typical).

IHI = number of inequality constraint multipliers HI that are to be initialized to some positive nonzero constant (number of $HI(J) > 0$ initially). IHI $\in [0,NI]$.
 = 0 if NI = 0 or if all equality constraint multipliers are to be zero initially (this is typical).

(DATA CARD 3, continued)

> IHVL = number of multipliers, associated with the lower bounded variables, that are to be initialized to some nonzero positive constant (number of HVL(K)>0). IHVL $\in$ [0,N].
> = 0 if NBV = 0 or if all lower-bound constraint multipliers are to be initialized to zero (initially if HVL(K) = 0 for all lower bounded X(K)).
>
> IHVU = number of multipliers, associated with the upper bounded variables, that are to be initialized to some nonzero positive constant (number of HVU(K)>0). IHVU $\in$ [0,N].
> = 0 if NBV = 0 or if all upper-bound constraint multipliers are to be initialized to zero (if initially HVU(K) = 0 for all upper bounded X(K)).

DATA CARD 4:

    READ 1010,EPS1,EPS2,EPS3
1010 FØRMAT (8D10.5)

> EPS1 = part of convergence criterion; LPNLP stops when GMAG < EPS1 and DELX < EPS3, where GMAG is the magnitude of the gradient of the augmented Lagrangian, and DELX is the magnitude change in x between consecutive search points. This test is made only at the end of a cycle, after the multipliers and weights have been updated. Typical values of EPS1 are (1.D-04 to 1.D-08).

A.3 INSTRUCTIONS FOR USING LPNLP

(DATA CARD 4, continued)

    EPS2,EPS3 = part of update criterion; multipliers and penalty weights are updated when DELX < EPS3 and GMAG < EPV where EPV is assigned the value of EPS2 initially and is dynamically reduced to EPS1 (see Chapter 5). EPS2 does not have to equal EPS3, but typical values are (0.01, 0.01) to (0.0001, 0.0001).

DATA CARD 5:

```
      READ 1010,(X(I),I=1,N)
1010  FØRMAT (8D10.5)
```

This card provides the initial x_i values. List 8 values per card using the D10.5 format field with as many cards as necessary to list all N values X(1), ..., X(N). The last card may have less than 8 values.

DATA CARD 6:

```
      NEPI = NE+NI+NBV
1010  FØRMAT (8D10.5)
      IF(NEPI.NE.0) READ 1010,W1,W2,W3,W1MAX,
     &W2MAX,W3MAX,WF
```

This card provides the initial penalty weights. If the problem is unconstrained (i.e., if NE+NI+NBV=0), this card must not be included. Otherwise, the penalty weights are initialized as follows:

(DATA CARD 6, continued)

$W1$ = initial weight for all equality constraints.
= 0 if $NE = 0$.

$W2$ = initial weight for all inequality constraints and bounded variables with associated positive valued multipliers.
= 0 if $NI+NBV=0$.

$W3$ = initial weight for all inequality constraints and bounded variables with associated zero valued multipliers.
= 0 if $NI+NBV=0$.

$W1MAX$ = maximum value for $W1$; $W1$ is held fixed at $W1MAX$ when $W1 \geq W1MAX$.
= 0 if $NE = 0$.

$W2MAX$ = maximum value for $W2$; $W2$ is held fixed at $W2MAX$ when $W2 \geq W2MAX$.
= 0 if $NI+NBV=0$.

$W3MAX$ = maximum value for $W3$; $W3$ is held fixed at $W3MAX$ when $W3 \geq W3MAX$.
= 0 if $NI+NBV=0$.

WF = factor by which $W1$, $W2$, and $W3$ are updated (increased) through multiplication by WF. Typically ($2.0 \leq WF \leq 4.0$).
= 1.0 if weights are to be held fixed at initial values.

For a scaled problem, usually $W1$, $W2$, and $W3$ are initialized in the range $(0.25, 2)$ and $W1MAX$, $W2MAX$, and $W3MAX$ are in the range $(8, 64)$. If a constraint breakthrough occurs, preventing

A.3 INSTRUCTIONS FOR USING LPNLP

convergence, initial and maximum values of all the above parameters except WF should be increased and the problem rerun.

Note: The index limits for the remaining cards are contained on Data Card 3.

DATA CARD(S) 7:

```
          IF (IGF.EQ.0) GØ TØ 20
          READ 1090,(K,GF(K),L=1,IGF)
     1090 FØRMAT (4(I5,D15.5))
       20 CØNTINUE
```

This card(s) gives the constant gradient components of the objective function. If IGF = 0, this card must not be included. If IGF ≠ 0, list the index K, and the constant value GF(K) for each component using the (I5,D15.5) format field. There are four such fields per card, allowing the input of up to 4 indices and constant gradient components per card. Use as many cards as necessary to list all IGF indices and components. The last card may have less than 4 fields. Constant gradient components equal to zero (GF(K) = 0) need not be read, as all gradient components are set to zero before this data card(s) is read or before subroutine GRAD is called.

Note: If NE = 0, Data Cards 8, 9, and 10 must not be included, otherwise they are as follows:

DATA CARD(S) 8:

```
     IF (NA.NE.0) READ 1090,(I,A(I),L=1,NA)
1090 FØRMAT (4(I5,D15.5))
```

This card(s) provides the nonzero constants for the right-hand sides of the equality constraints. This card(s) must not be included if NA = 0. If NA ≠ 0, list the index I, and the value A(I) for each constant using the (I5,D15.5) format field. There are 4 such fields per card allowing up to 4 indices and constants per card. Use as many cards as necessary to list all NA indices and constants. The last card may have less than 4 fields. Constants equal to zero (A(I)=0) need not be read as all A(I) values are set to zero before this data card(s) is read.

A.3 INSTRUCTIONS FOR USING LPNLP

DATA CARD(S) 9:

```
      IF (IHE.NE.0) READ 1090,(I,HE(I),L=1,IHE)
1090  FØRMAT (4(I5,D15.5))
```

This card(s) gives the nonzero initial values for the equality constraint multipliers. This card must not be included if IHE = 0. If IHE ≠ 0, list the index I, and value HE(I) for each multiplier using the (I5,D15.5) format field. There are 4 such fields per card allowing up to 4 indices and multiplier values per card. Use as many cards as necessary to list all IHE indices and multiplier values. The last card may have less than 4 fields. Multipliers equal to zero (HE(I)=0) need not be read as all HE(I) values are set to zero before this data card(s) is read.

DATA CARD(S) 10:

```
      IF (IGE.EQ.0) GØ TØ 70
      READ 1090,(I,GE(I),L=1,IGE)
1090  FØRMAT (4(I5,D15.5))
  70  CØNTINUE
```

This card(s) supplies the constant gradient components for the equality constraint functions. This card(s) must not be included if IGE = 0. If

IGE ≠ 0, the constant gradient components are read with the same format specifications as Data Card 7.

Note: If NI = 0, Data Cards 11, 12, and 13 must not be included, otherwise they are as follows:

DATA CARD(S) 11:

```
      IF (NB.NE.0) READ 1090,(J,B(J),L=1,NB)
 1090 FØRMAT (4(I5,D15.5))
```

This card(s) provides the nonzero constants for the right-hand sides of the inequality constraints. This card(s) must not be included if NB = 0. If NB ≠ 0, then the B(J) constants are read with the same format specifications as Data Card 8.

DATA CARD(S) 12:

```
      IF (IHI.NE.0) READ 1090,(J,HI(J),L=1,IHI)
 1090 FØRMAT (4(I5,D15.5))
```

This card(s) provides the nonzero positive initial values for the inequality constraint

A.3 INSTRUCTIONS FOR USING LPNLP 313

multipliers. This card must not be included if IHI = 0. If IHI ≠ 0, the nonzero positive multipliers are read with the same format specifications as Data Card(s) 9.

DATA CARD(S) 13:

```
      IF (IGI.EQ.0) GØ TØ 120
      READ 1090,(J,GI(J),L=1,IGI)
1090  FØRMAT (4(I5,D15.5)
 120  CØNTINUE
```

This card(s) provides the constant gradient components for the inequality constraint functions. This card(s) must not be included if IGI = 0. If IGI ≠ 0, the constant gradient components are read with the same format specifications as Data Card(s) 10.

Note: If NBV = 0, Data Cards 14, 15, and 16 must not be included, otherwise they are as follows:

DATA CARD(S) 14:

```
      IF (NBV.EQ.0) GØ TØ 180
      READ 1100,(K,IX(K),C(K),D(K),L=1,NBV)
 1100 FØRMAT (2(I5,I5,D15.5,D15.5))
  180 CØNTINUE
```

This card(s) provides the upper and lower bounds C(K) and D(K) for the X(K) variables. This card must not be included if NBV = 0. If NBV $\neq$ 0, then the index K, the integer code IX(K), the lower bound C(K), and the upper bound D(K) are read for each bounded variable X(K) using the (I5,I5,D15.5,D15.5) format field. The integer code is as follows:

$$IX(K) = \begin{cases} 0 & \text{if } X(K) \text{ is unbounded} \\ 1 & \text{if } X(K) \text{ is bounded from below only} \\ 2 & \text{if } X(K) \text{ is bounded from above only} \\ 3 & \text{if } X(K) \text{ is bounded from both below and above} \end{cases}$$

The components of the IX vector are all set to zero before this card(s) is read. Therefore, only the bounded variables need be read (having a code of 1, 2, or 3). If IX(K) has a code 1, then K, IX(K), C(K), ..., are listed on the card with the D(K) field left blank. If IX(K) has a code 2, then K, IX(K), ..., D(K) are listed on the card

A.3 INSTRUCTIONS FOR USING LPNLP

with the C(K) field left blank. If IX(K) has a code 3, then K, IX(K), C(K), D(K) are listed on the card. There are 2 such fields per card allowing the input of the index, integer code, lower bound, and upper bound for 2 variables per card. Use as many cards as necessary to list all NBV bound sets. The last card may contain information for only one variable.

DATA CARD(S) 15:

```
      IF (IHVL.NE.0) READ 1090,(K,HVL(K),L=1,IHVL)
1090  FØRMAT (4(I5,D15.5))
```

This card provides the nonzero positive initial values for the multipliers associated with the lower bounded X(K) variables. This card must not be included if IHVL = 0. If IHVL ≠ 0, the nonzero positive multipliers are read with the same format specifications as Data Card(s) 9.

DATA CARD(S) 16:

```
      IF (IHVU.NE.0) READ 1090,(K,HVU(K),L=1,IHVU)
1090  FØRMAT (4(I5,D15.5))
```

(DATA CARD 16, continued)

This card(s) provides the nonzero positive initial values for the multipliers associated with the upper bounded X(K) variables. This card(s) must not be included if IHVU = 0. If IHVU ≠ 0, the nonzero positive multipliers are read with the same format specifications as Data Card(s) 9.

STACKING DATA DECKS. Data decks may be stacked for as many runs on the <u>same</u> problem as desired, each with different initial conditions, different bounds on variables, or different right-hand-side constants in constraints. Each data deck must contain the correct order and number of data cards as described above. To stack data, a loop must be introduced into the main calling program such that LPNLP is called once for each data deck included.

A.4 CONTROL CARDS

Control cards tend to be computer-installation dependent; and therefore, the following set provides an example of what may be necessary. The system control cards for the XDS Sigma 7 computer

A.4 INSTRUCTIONS FOR USING LPNLP

at Montana State University for any problem to be solved by LPNLP are

```
!JØB account #, name
!LIMIT (TIME,5),(other limits as may be required)
!FØRTRAN LS,GØ,ADP
            user supplied routines
            (calling routine, FXN, and GRAD)
!LØAD (GØ),(EF,(NLPB))
!RUN
!DATA
            data deck 1
            data deck 2
            etc.
!EØD
```

The !LØAD card indicates, among other things, that a compiled version NLPB of the set of subroutines that constitute the general routine LPNLP is to be found in mass storage under the given account number.

Note: LPNLP is written for single precision calculations to work on most computers. The XDS Sigma 7 at Montana State University has a control statement for "automatic double precision" (ADP) as indicated on the !FØRTRAN card. If this program is to be run on a different computer system, modifications must be made to comply with that system for double precision, which may mean modifications to the subroutines contained in LPNLP. It is advisable to run under double precision. Problems may be run under single precision, but convergence may be slower, or may not be obtained at all due to the numerical accuracy required for the unidirectional search to be efficient as the optimum value of x is approached.

If modifications must be made to the subroutines, insertion of the specification statement

```
IMPLICIT REAL*8 (A-H,Ø-Z)
```

into each subroutine will allow for the algorithm to operate in double precision.

To better illustrate the requirements for proper implementation of the subroutines and data, one sample problem is given (Test Problem T9 of Chapter 6) in the next section, showing proper user supplied routines and data.

A.5 SAMPLE CASE: TEST PROBLEM T9 OF CHAPTER 6.

```
      SUBROUTINE FXNS(X,F,FE,FI)
      DIMENSION X(1),FE(1),FI(1)
      F = 5.*X(1) + 5.*X(2) + 4.*X(3)+ X(1)*X(3) + 6.*X(4)
     & + 5.*X(5)/(1.+X(5)) + 8.*X(6)/(1.+X(6))
     & + (10. -20.*EXP(-X(7))+10.*EXP(-2.*X(7)))
      FE(1) = 2.*X(4) + X(5) + .8*X(6) + X(7)
      FE(2) = X(2)**2 + X(3)**2 + X(5)**2 + X(6)**2
      FI(1) = X(1) + X(2) + X(3) + X(4) + X(5) + X(6) + X(7)
      FI(2) = X(1) + X(2) + X(3) + X(4)
      FI(3) = X(1) + X(3) + X(5) + X(6)**2 - X(7)**2
      RETURN
      END
      SUBROUTINE GRAD(X,GF,GE,GI)
      DIMENSION X(1),GF(1),GE(1),GI(1)
      GF(1) = 5. + X(3)
      GF(3) = 4. + X(1)
      GF(5) = 5./(1.+X(5))**2
      GF(6) = 8./(1.+X(6))**2
      GF(7) = 20.*EXP(-X(7)) - 20.*EXP(-2.*X(7))
      GE(9) = 2.*X(2)
      GE(10) = 2.*X(3)
      GE(12) = 2.*X(5)
      GE(13) = 2.*X(6)
      GI(20) = 2.*X(6)
      GI(21) = -2.*X(7)
      RETURN
      END
```

320   MATH PROGRAMMING VIA AUGMENTED LAGRANGIANS

(Calling Program)

```
C     SEVEN VARIABLE PROBLEM
C
      DIMENSION A(2),B(3),C(7),D(7),DLX(7),DLG(7),FE(2),FES(2)
      DIMENSION FI(3),FIS(3),GAL(7),GE(14),GF(7)
      DIMENSION GI(21),GO(7),H(49),HE(2),HI(3),HVL(7),HVU(7),IX(7)
      DIMENSION R(7),X(7),XT(7)
      CALL LPNLF(A,B,C,D,DLX,DLG,FE,FES,FI,FIS,GAL,GE,GF,GI,GO,
     &H,HE,HI,HVL,HVU,IX,R,X,XT)
      STOP
      END
```

(Data)

```
SEVEN VARIABLE PROBLEM
7    2    3    7   100   20    0   15    0    0            LPNLP
2    3    2    4    14
.0001    .1        .1        .1        .1        .1
.1       .1        .1        32.       32.       32.       4.
1.       1.        1.
2.       5.        4.        6.                                      7.   1.
1.       5.        2.        5.                  .8
4.       2.        5.        1.        6.        5.                  4.   1.
1.      10.        2.        5.        3.        1.                  8.   1.
1.       1.        2.        1.        3.        1.                 15.   1.
5.       1.        6.        1.        7.        1.
9.       1.       10.        1.       11.        1.
17.      1.       19.        1.
1.       1.                            2.        1.
3.       1.                            4.        1.
5.       1.                            6.        1.
7.       1.
```

A.5 INSTRUCTIONS FOR USING LPNLP   321

(Initial-Value Output)

```
SEVEN VARIABLE PROBLEM                                    LPNLP
################### INITIAL VALUES ####################################

   N =  7       NE =  2      NI =  3       NBV =  7

  NA    NB    IGF   IGE   IGI   IHE   IHI   IHVL   IHVU
   2     3     2     4    14     0     0     0      0

EPS1 = .100D-03        EPS2 = .100D 00        EPS3 = .100D 00

MAXIMUM ITERATION LIMIT... 100 ; OUTPUT PRINTED EVERY 20 ITERATION(S)
WEIGHT/MULTIPLIER UPDATE INFORMATION IS PRINTED

  15 IS THE MAXIMUM NUMBER OF UNIDIRECTIONAL SEARCHES DURING EACH
MAXIMIZATION PHASE (MULTIPLIERS AND WEIGHTS ARE HELD CONSTANT WITHIN
EACH SUCH PHASE).

ISS = 0         IRESET = 0

INITIAL PENALTY WEIGHTS ARE         1.00  1.00     1.00
RESPECTIVELY, AND ARE UPDATED BY A FACTOR OF       4.00 UNTIL THEY EXCEED
VALUES OF  .320D 02  .320D 02  .320D 02

CONSTANT GRADIENT COMPONENTS OF THE FUNCTION(S)
   2  .500000D 01   4  .600000D 01

EQUALITY CONSTRAINT STATUS:
  I     FE(I)          A(I)          A(I)-FE(I)         HE(I)
  1   .480000D 00   .500000D 01     .452000D 01      .000000D 00
  2   .400000D-01   .500000D 01     .496000D 01      .000000D 00
```

(Initial-Value Output, continued)

```
CONSTANT GRADIENT COMPONENTS OF THE FUNCTION(S)
     4   .200000D 01    5   .100000D 01    6   .800000D 00    7   .100000D 01

INEQUALITY CONSTRAINT STATUS:
     J       FI(J)            B(J)           B(J)-FI(J)          HI(J)
     1    .700000D 00     .100000D 02       .930000D 01       .000000D 00
     2    .400000D 00     .500000D 01       .460000D 01       .000000D 00
     3    .300000D 00     .500000D 01       .470000D 01       .000000D 00

CONSTANT GRADIENT COMPONENTS OF THE FUNCTION(S)
     1   .100000D 01    2   .100000D 01    3   .100000D 01    4   .100000D 01
     5   .100000D 01    6   .100000D 01    7   .100000D 01
     8   .100000D 01    9   .100000D 01   10   .100000D 01   11   .100000D 01
    15   .100000D 01   17   .100000D 01   19   .100000D 01

CONSTRAINED VARIABLE STATUS:
   HVL(K)          C(K)            K         D(K)            HVU(K)
  .000000D 00    .000000D 00       1
  .000000D 00    .000000D 00       2
  .000000D 00    .000000D 00       3
  .000000D 00    .000000D 00       4
  .000000D 00    .000000D 00       5
  .000000D 00    .000000D 00       6
  .000000D 00    .000000D 00       7

X VALUES, X(1),...,X(N)
  .100000D 00    .100000D 00    .100000D 00    .100000D 00
  .100000D 00    .100000D 00

### F = .328238D 01   AUG. LAG. = -.417496D 02   GMAG = .358845D 02 ##
####################################################################
```

A.5 INSTRUCTIONS FOR USING LPNLP

(Output for Typical Update Phase)

```
###############################################################
## MULTIPLIERS AND PENALTY WEIGHTS UPDATED AT SEARCH POINT *  15 **

EQUALITY CONSTRAINT VALUES, FE(1),...,FE(NE)
 .475273D 01  .529245D 01

EQUALITY CONSTRAINT MULTIPLIERS, HE(1),...,HE(NE)
-.494546D 00  .584896D 00

INEQUALITY CONSTRAINT VALUES, FI(1),...,FI(NI)
 .110685D 02  .743663D 01  .279403D 01

INEQUALITY CONSTRAINT MULTIPLIERS, HI(1),...,HI(NI)
 .213701D 01  .487326D 01  .000000D 00

              LOWER                BOUNDED X(K) VARIABLES        UPPER
   HVL(K)           X(K)-C(K)      K       D(K)-X(K)           HVU(K)
 .000000D 00        .541513D 01    1
 .126905D 01       -.634527D 00    2
 .000000D 00        .202170D 01    3
 .000000D 00        .634322D 00    4
 .000000D 00        .506415D 00    5
 .000000D 00        .738979D 00    6
 .000000D 00        .238648D 01    7

W1 =  .400000D 01   W2 =  .400000D 01   W3 =  .400000D 01
$$$ F =  .600695D 02   AUG. LAG. =  .143005D 02   GMAG =  .583179D 02
###############################################################
```

(Search-Point Output)

```
*************************** SEARCH POINT ***    20 *******************************
FXNS CALLS ...      48      GRAD CALLS ...      21      RHO =    .632705D 00
$$$ F =   .446533D 02   AUG. LAG. =   .445116D 02   GMAG =   .953384D 00
SEARCH DIRECTION TO THIS POINT, R(1),...,R(N)                    NG =   20
  .342866D-01  -.728839D-02  -.294627D-01  -.462219D-01   .291136D-01
 -.320040D-01  -.231475D-01
DELX =   .515601D-01           DELG =   .109988D 01          RMAG =   .814916D-01
X VALUES, X(1),...,X(N)
  .319675D 01  -.492574D-01   .171886D 01   .178646D 00   .827197D 00
  .115502D 01   .289291D 01
############### CONVERGENCE CRITERION SATISFIED ###############################
              FINAL SEARCH POINT VALUES TO FOLLOW
```

(Final Values)

```
*************************** SEARCH POINT ***    37 *******************************
FXNS CALLS ...      78      GRAD CALLS ...      38      RHO =    .100000D 01
$$$ F =   .444687D 02   AUG. LAG. =   .444687D 02   GMAG =   .152191D-04
SEARCH DIRECTION TO THIS POINT, R(1),...,R(N)                    NG =   37
  .774062D-05  -.201373D-05  -.300414D-05  -.871322D-05   .273354D-05
  .158689D-05   .115028D-04
DELX =   .170652D-04           DELG =   .799457D-03          RMAG =   .170652D-04
```

A.5 INSTRUCTIONS FOR USING LPNLP

(Final Values, continued)

```
X VALUES, X(1),...,X(N)
  .324182D 01    .630303D-07    .163416D 01    .124021D 00    .889614D 00
  .124021D 01    .287018D 01

EQUALITY CONSTRAINT STATUS:
   I        FE(I)           A(I)          A(I)-FE(I)         HE(I)
   1     .500000D 01    .500000D 01     -.284185D-07      -.317079D 00
   2     .500000D 01    .500000D 01     -.588209D-07       .185926D 00

INEQUALITY CONSTRAINT STATUS:
   J        FI(J)           B(J)          B(J)-FI(J)         HI(J)
   1     .100000D 02    .100000D 02      .205609D-06       .138658D 01
   2     .500000D 01    .500000D 01     -.149028D-06       .524757D 01
   3    -.934214D 00    .500000D 01      .593421D 01       .000000D 00

CONSTRAINED VARIABLE STATUS:
         LOWER         BOUNDED X(K)  VARIABLES         UPPER
        HVL(K)        X(K)-C(K)        K        D(K)-X(K)        HVU(K)
      .000000D 00    .324182D 01      1
      .163416D 01    .630303D-07      2
      .000000D 00    .163416D 01      3
      .000000D 00    .124021D 00      4
      .000000D 00    .889614D 00      5
      .000000D 00    .124021D 01      6
      .000000D 00    .287018D 01      7

  M = .320000D 02   W2 = .320000D 02   W3 = .320000D 02
```

APPENDIX B

B FORTRAN SUBROUTINES

This appendix contains the 13 general subroutines of the multiplier algorithm described in Chapter 5. These subroutines are used in conjunction with a calling program and with the problem dependent subroutines FXNS and GRAD (see Appendix A). The hierarchy of the subroutines is implicit in the following list.

    Calling Program calls LPNLP.

    LPNLP   calls   INITL, DFPRV, SEARCH, GRAD, GALAG, DELTA, UPDATE, AUGLAG, and OUTPUT.

    INITL   calls   CONGR, FXNS, GRAD, AUGLAG, and GALAG.

    SEARCH calls   VALUE and DMAX.

    VALUE   calls   FXNS and AUGLAG.

B FORTRAN SUBROUTINES 327

```
      SUBROUTINE LFNLF(A,B,C,D,DLX,DLG,FE,FES,FI,FIS,GAL,GE,GF,GI,     LNLF0010
     &GO,H,HE,HI,HVL,HVU,IX,R,X,XT)                                    LNLF0020
C                                                                     LNLF0030
C  COMMAND PROGRAM; ORGANIZATION AND OPERATION OF ALGORITHM           LNLF0040
C                                                                     LNLF0050
      DIMENSION A(1),B(1),C(1),D(1),DLX(1),DLG(1),FE(1),FES(1),FI(1)   LNLF0060
      DIMENSION FIS(1),GAL(1),GE(1),GF(1),GI(1),GO(1),X(1),XT(1)       LNLF0070
      DIMENSION H(1),HE(1),HI(1),HVL(1),HVU(1),IX(1),R(1)              LNLF0080
      COMMON /C1/N,NE,NI,NEV,NEPI/C2/NFE,NGE/C3/W1,W2,W3               LNLF0090
      COMMON /C4/W1MAX,W2MAX,W3MAX,WF/C6/DELX,DELG                     LNLF0100
      COMMON /C9/KSRCH,IBSD/C20/NUF                                    LNLF0110
      COMMON /C10/EPS1,EPS2,EPS3/C11/ISTOP,ITOT                        LNLF0120
      COMMON /C12/NSRCH,IT,NFRINT,IMAX                                 LNLF0130
      COMMON /C13/RMAG,RHO,GMAG,SLOPE/C21/F,AL                         LNLF0140
      COMMON /C17/DS,FS,YS/C18/NG,IUF,IFAIL                            LNLF0150
      COMMON /C25/ISS,IRESET,NN                                        LNLF0160
C                                                                     LNLF0170
C  INITIALIZATION                                                     LNLF0180
C                                                                     LNLF0190
      CALL INITL(A,B,C,D,FE,FI,GAL,GE,GF,GI,GO,HE,HI,HVL,HVU,IX,X)     LNLF0200
      ICONV=0                                                          LNLF0210
      IFAIL=0                                                          LNLF0220
      NED=0                                                            LNLF0230
      EPD=0.1*EPS1                                                     LNLF0240
      EFT=10.*EPS2                                                     LNLF0250
      EFV=EPS2                                                         LNLF0260
      IF(NE.EQ.0) GO TO 3                                              LNLF0270
```

```
      DO 2 I=1,NE                                                       LNLF0280
    2 FES(I)=FE(I)                                                      LNLF0290
    3 IF(NI.EQ.0) GO TO 5                                               LNLF0300
      DO 4 I=1,NI                                                       LNLF0310
    4 FIS(I)=FI(I)                                                      LNLF0320
    5 FS=F                                                              LNLF0330
C                                                                       LNLF0340
C     GENERATE SEARCH DIRECTION R; CONDUCT UNIDIRECTIONAL SEARCH;       LNLF0350
C     AND UPDATE SEARCH-POINT STATUS                                    LNLF0360
C                                                                       LNLF0370
   10 IF(ITOT.EQ.IMAX) GO TO 80                                         LNLF0380
      CALL DFFRV(GAL,H,DLX,DLG,N,R)                                     LNLF0390
      IF(NG-2) 11,12,14                                                 LNLF0400
   11 NBD=NBD+1                                                         LNLF0410
      IF(NBD-5) 14,80,80                                                LNLF0420
   12 NBD=0                                                             LNLF0430
   14 CALL SEARCH(A,B,C,D,FE,FES,FI,FIS,HE,HI,HVL,HVU,IX,               LNLF0440
     &X,XT,R)                                                           LNLF0450
      IF(RHO.EQ.0.) GO TO 15                                            LNLF0460
      CALL GRAD(XT,GF,GE,GI)                                            LNLF0470
      NGE=NGE + 1                                                       LNLF0480
      CALL GALAG(A,B,C,D,FE,FI,GAL,GE,GF,GI,HE,HI,HVL,HVU,IX,XT)        LNLF0490
   15 CALL DELTA(XT,X,GAL,GO,DLX,DLG,N)                                 LNLF0500
C                                                                       LNLF0510
C     INCREMENT COUNTERS; TEST FOR OUTPUT, AND TEST FOR UPDATE          LNLF0520
C                                                                       LNLF0530
      ITOT=ITOT + 1                                                     LNLF0540
      KSRCH=KSRCH + 1                                                   LNLF0550
      IT=IT + 1                                                         LNLF0560
      IF(IT.EQ.NPRINT.OR.RHO.EQ.0.) GO TO 110                           LNLF0570
```

```
   20 IF (IUP.EQ.0) GO TO 24                                          LNLF0580
      IFAIL=0                                                         LNLF0590
      IF(DELX.GT.EF$3) GO TO 28                                       LNLF0600
      IF(GMAG.LT.EPV) GO TO 40                                        LNLF0610
      IF(DELG.GT.EPV) GO TO 28                                        LNLF0620
      IF(NG.NE.1) GO TO 26                                            LNLF0630
      IF(DELG.LT.EPD.AND.KSRCH.EQ.1) GO TO 80                         LNLF0640
      GO TO 26                                                        LNLF0650
   24 IFAIL=IFAIL+1                                                   LNLF0660
      IF(IFAIL.EQ.4) GO TO 80                                         LNLF0670
      IF(NG.EQ.1 .OR. RHO.EQ.0.) GO TO 40                             LNLF0680
   26 IBSD=IBSD+1                                                     LNLF0690
   28 IF(KSRCH.GT.N) GO TO 30                                         LNLF0700
      IF(IBSD.EQ.2) KSRCH=N                                           LNLF0710
      GO TO 10                                                        LNLF0720
   30 IF(KSRCH.EQ.NSRCH) GO TO 40                                     LNLF0730
      IF(GMAG-EPV) 40,40,35                                           LNLF0740
   35 IF(IBSD.LT.3) GO TO 10                                          LNLF0750
      RHO=0.                                                          LNLF0760
C                                                                     LNLF0770
C     UPDATE MULTIPLIERS AND PENALTY WEIGHTS                          LNLF0780
C                                                                     LNLF0790
   40 IF(NEPI.EQ.0) GO TO 70                                          LNLF0800
      IF(NUP.EQ.0) PRINT 1000,ITOT                                    LNLF0810
      CALL UPDATE(A,B,C,D,FE,FI,HE,HI,HVL,HVU,IX,X)                   LNLF0820
      CALL AUGLAG(A,B,C,D,FE,FI,HE,HI,HVL,HVU,IX,X,F,AL)              LNLF0830
      CALL GALAG(A,B,C,D,FE,FI,GAL,GE,GF,GI,HE,HI,HVL,HVU,IX,X)       LNLF0840
      GMAG=0.                                                         LNLF0850
```

```
      DO 60 I=1,N
      GO(I)=GAL(I)
   60 GMAG=GMAG + GAL(I)*GAL(I)
      GMAG=SQRT(GMAG)
      IF(NUF.EQ.0) PRINT 1010,F,AL,GMAG
   70 KSRCH=0
      IBSD=0
      IF(GMAG.LT.EPT)    EPV=EPS1
C
C     TEST:  CONVERGENCE CRITERION
C
      IF(GMAG.GT.EPS1.OR.DELX.GT.EPS3) GO TO 10
   75 ICONV=1
C
C     STOP:  CONVERGENCE OR ITERATION LIMIT
C     ..FINAL SEARCH POINT INFORMATION..
C
   80 ISTOP=1
      PRINT 1050
      IF(ICONV.EQ.1) GO TO 90
      IF(IFAIL.EQ.4) GO TO 95
      IF(ITOT.NE.IMAX) GO TO 85
      PRINT 1020
      GO TO 100
   85 PRINT 1070
      PRINT 1080
      GO TO 100
   90 PRINT 1030
      GO TO 100
   95 PRINT 1060
```

B FORTRAN SUBROUTINES 331

```
      PRINT 1080                                                        LNLF1160
  100 PRINT 1040                                                        LNLF1170
  110 CALL OUTPUT(A,B,C,D,FE,FI,HE,HI,HVL,HVU,IX,X,R)                   LNLF1180
      IT=0                                                              LNLF1190
      IF(ISTOP.EQ.0) GO TO 20                                           LNLF1200
      RETURN                                                            LNLF1210
C                                                                       LNLF1220
C   ********** FORMATS **********                                       LNLF1230
C                                                                       LNLF1240
 1000 FORMAT(//,20X,'####################################',             LNLF1250
     &'###################',//,20X,' MULTIPLIERS AND PENALTY',          LNLF1260
     &' WEIGHTS UPDATED AT SEARCH POINT ',I6,' **')                     LNLF1270
 1010 FORMAT(/,20X,'$$$ F = ',D13.6,3X,'AUG. LAG. = ',D13.6,3X,         LNLF1280
     &'GMAG = ',D13.6,//,20X,'####################################',    LNLF1290
     &'###################')                                            LNLF1300
 1020 FORMAT(20X,'####################### MAXIMUM ITERATION LIMIT ',    LNLF1310
     &'####################')                                           LNLF1320
 1030 FORMAT(20X,'######################## CONVERGENCE CRITERION ',     LNLF1330
     &'SATISFIED ####################')                                 LNLF1340
 1040 FORMAT(20X,'............. FINAL SEARCH POINT VALUES TO FOLLOW',   LNLF1350
     &'..............')                                                 LNLF1360
 1050 FORMAT('1')                                                       LNLF1370
 1060 FORMAT(20X,'######## A LARGER VALUE OF THE AUGMENTED LA',         LNLF1380
     &'GRANGIAN COULD NOT BE',//,20X,'FOUND IN 4 CONSECUTIVE SEARCHES:') LNLF1390
 1070 FORMAT(20X,'LACK OF ACCURACY OR POOR CONVERGENCE RATE DETECTED:') LNLF1400
 1080 FORMAT(20X,'YOU MAY WANT TO RERUN THE PROBLEM WITH A DIFFERENT ', LNLF1410
     &'STARTING POINT',//,20X,'AND/OR DIFFERENT SEARCH PARAMETERS AND/', LNLF1420
     &'OR A DIFFERENT MODE OF SEARCH--',//,20X,'DIRECTION GENERATION.',/) LNLF1430
      END                                                               LNLF1440
```

```
      SUBROUTINE INITL(A,B,C,D,FE,FI,GAL,GE,GF,GI,GO,HE,HI,HVL,HVU,IX,X)     INTL0010
C                                                                            INTL0020
C     READS DATA AND INITIALIZES ALGORITHM; OUTPUTS INITIAL CONDITIONS       INTL0030
C                                                                            INTL0040
      DIMENSION A(1),B(1),C(1),D(1),FE(1),FI(1),GAL(1),GE(1),GF(1)           INTL0050
      DIMENSION GI(1),GO(1),HE(1),HI(1),HVL(1),HVU(1),IX(1),X(1),ID(18)      INTL0060
      COMMON /C1/N,NE,NI,NEV,NEFI/C2/NFE,NGE/C3/W1,W2,W3                     INTL0070
      COMMON /C4/W1MAX,W2MAX,W3MAX,WF/C9/KSRCH,IBSD/C20/NUF                  INTL0080
      COMMON /C10/EPS1,EPS2,EPS3/C11/ISTOP,ITOT/C21/F,AL                     INTL0090
      COMMON /C12/NSRCH,IT,NPRINT,IMAX/C13/RMAG,RHO,GMAG,SLOPE               INTL0100
      COMMON /C25/ISS,IRESET,NN                                              INTL0110
C                                                                            INTL0120
C     INITIAL CONDITIONS                                                     INTL0130
C                                                                            INTL0140
      RHO=0.                                                                 INTL0150
      GMAG=0.                                                                INTL0160
      IBSD=0                                                                 INTL0170
      ISTOP=0                                                                INTL0180
      IT=0                                                                   INTL0190
      ITOT=0                                                                 INTL0200
      KSRCH=0                                                                INTL0210
      NFE=1                                                                  INTL0220
      NGE=1                                                                  INTL0230
      READ 1170,(ID(I),I=1,18)                                               INTL0240
      READ 1000,N,NE,NI,NEV,IMAX,NPRINT,NUF,NSRCH,ISS,IRESET                 INTL0250
      NN=N*N                                                                 INTL0260
```

B FORTRAN SUBROUTINES

```
      READ 1000,NA,NB,IGF,IGE,IGI,IHE,IHI,IHVL,IHVU              INTL0270
      READ 1010,EPS1,EPS2,EPS3                                    INTL0280
      READ 1010,(X(I),I=1,N)                                      INTL0290
      NEPI=NE+NI+NEV                                              INTL0300
      IF(NEPI.NE.0) READ 1010,W1,W2,W3,W1MAX,W2MAX,W3MAX,WF       INTL0310
      PRINT 1160                                                  INTL0320
      PRINT 1180,(ID(I),I=1,18)                                   INTL0330
      PRINT 1020                                                  INTL0340
      PRINT 1030,N,NE,NI,NEV                                      INTL0350
      PRINT 1300,NA,NB,IGF,IGE,IGI,IHE,IHI,IHVL,IHVU              INTL0360
      PRINT 1040,EPS1,EPS2,EPS3                                   INTL0370
      PRINT 1050,IMAX,NPRINT                                      INTL0380
      IF(NUP.EQ.1) PRINT 1210                                     INTL0390
      IF(NUP.EQ.0) PRINT 1220                                     INTL0400
      PRINT 1060,NSRCH                                            INTL0410
      PRINT 1310,ISS,IRESET                                       INTL0420
      IF(NEPI.EQ.0) GO TO 5                                       INTL0430
      PRINT 1070,W1,W2,W3,WF,W1MAX,W2MAX,W3MAX                    INTL0440
C                                                                 INTL0450
C     COST FUNCTION INITIALIZATION                                INTL0460
C                                                                 INTL0470
    5 DO 10 K=1,N                                                 INTL0480
   10 GF(K)=0.                                                    INTL0490
      IF(IGF.EQ.0) GO TO 20                                       INTL0500
      READ 1090,(K,GF(K),L=1,IGF)                                 INTL0510
      CALL CONGR(GF,GO,IX,N,1)                                    INTL0520
   20 CALL FXNS(X,F,FE,FI)                                        INTL0530
C                                                                 INTL0540
```

```
      C     EQUALITY CONSTRAINT INITIALIZATION                  INTL0550
      C                                                         INTL0560
            PRINT 1190                                          INTL0570
            IF(NE.EQ.0) GO TO 60                                INTL0580
            DO 30 I=1,NE                                        INTL0590
            A(I)=0.                                             INTL0600
         30 HE(I)=0.                                            INTL0610
            K=N*NE                                              INTL0620
            DO 40 I=1,K                                         INTL0630
         40 GE(I)=0.                                            INTL0640
            IF(NA.NE.0) READ 1090,(I,A(I),L=1,NA)                INTL0650
            IF(IHE.NE.0) READ 1090,(I,HE(I),L=1,IHE)            INTL0660
            PRINT 1200                                          INTL0670
            DO 50 I=1,NE                                        INTL0680
            TEMP=A(I)-FE(I)                                     INTL0690
         50 PRINT 1150,I,FE(I),A(I),TEMP,HE(I)                  INTL0700
            IF(IGE.EQ.0) GO TO 70                               INTL0710
            READ 1090,(I,GE(I),L=1,IGE)                         INTL0720
            CALL CONGR(GE,GO,IX,N,NE)                           INTL0730
            GO TO 70                                            INTL0740
         60 PRINT 1140                                          INTL0750
      C                                                         INTL0760
      C     INEQUALITY CONSTRAINT INITIALIZATION                INTL0770
      C                                                         INTL0780
         70 PRINT 1230                                          INTL0790
            IF(NI.EQ.0) GO TO 110                               INTL0800
            DO 80 J=1,NI                                        INTL0810
            B(J)=0.                                             INTL0820
         80 HI(J)=0.                                            INTL0830
            K=N*NI                                              INTL0840
```

B FORTRAN SUBROUTINES

```
         DO 90 J=1,K                                              INTL0850
  90     GI(J)=0.                                                 INTL0860
         IF(NB.NE.0) READ 1090,(J,B(J),L=1,NB)                    INTL0870
         IF(IHI.NE.0) READ 1090,(J,HI(J),L=1,IHI)                 INTL0880
         PRINT 1240                                               INTL0890
         DO 100 J=1,NI                                            INTL0900
         TEMP=B(J)-FI(J)                                          INTL0910
 100     PRINT 1150,J,FI(J),B(J),TEMP,HI(J)                       INTL0920
         IF(IGI.EQ.0) GO TO 120                                   INTL0930
         READ 1090,(J,GI(J),L=1,IGI)                              INTL0940
         CALL CONGR(GI,GO,IX,N,NI)                                INTL0950
         GO TO 120                                                INTL0960
 110     PRINT 1140                                               INTL0970
C                                                                 INTL0980
C        BOUNDED VARIABLE INITIALIZATION                          INTL0990
C                                                                 INTL1000
 120     PRINT 1250                                               INTL1010
         IF(NBV.EQ.0) GO TO 180                                   INTL1020
         DO 130 K=1,N                                             INTL1030
         IX(K)=0                                                  INTL1040
         C(K)=0.                                                  INTL1050
         D(K)=0.                                                  INTL1060
         HVL(K)=0.                                                INTL1070
 130     HVU(K)=0.                                                INTL1080
         READ 1100,(K,IX(K),C(K),D(K),L=1,NBV)                    INTL1090
         IF(IHVL.NE.0) READ 1090,(K,HVL(K),L=1,IHVL)              INTL1100
         IF(IHVU.NE.0) READ 1090,(K,HVU(K),L=1,IHVU)              INTL1110
         PRINT 1260                                               INTL1120
```

```
      DO 170 K=1,N
      IVAR=IX(K)+1
      GO TO (170,140,150,160),IVAR
  140 PRINT 1270,HVL(K),C(K),K
      GO TO 170
  150 PRINT 1280,K,D(K),HVU(K)
      GO TO 170
  160 PRINT 1290,HVL(K),C(K),K,D(K),HVU(K)
  170 CONTINUE
      GO TO 190
  180 PRINT 1140
C
C     FORMULATE AUGMENTED LAGRANGIAN AND GRADIENT
C
  190 CALL GRAD(X,GF,GE,GI)
      CALL AUGLAG(A,B,C,D,FE,FI,HE,HI,HVL,HVU,IX,X,F,AL)
      CALL GALAG(A,B,C,D,FE,FI,GAL,GE,GF,GI,HE,HI,HVL,HVU,IX,X)
      DO 200 K=1,N
      GO(K)=GAL(K)
  200 GMAG=GMAG+GAL(K)*GAL(K)
      GMAG=SQRT(GMAG)
      PRINT 1110
      PRINT 1130,(X(I),I=1,N)
      PRINT 1120,F,AL,GMAG
      PRINT 1160
      RETURN
C
C     ********  FORMATS  *********
C
 1000 FORMAT(1015)
```

B   FORTRAN SUBROUTINES

```
1010 FORMAT(8D10.5)                                                     INTL1430
1020 FORMAT(//,20X,'############### INITIAL VALUES ###############     INTL1440
    &###########################')                                      INTL1450
1030 FORMAT(/,20X,4X,'N = ',I5,8X,'NE = ',I5,8X,'NI = ',I5,8X,          INTL1460
    &'NEV = ',I5)                                                       INTL1470
1040 FORMAT(20X,'EPS1 = ',D10.3,10X,'EPS2 = ',D10.3,10X,'EPS3 = ',      INTL1480
    &D10.3)                                                             INTL1490
1050 FORMAT(/,20X,'MAXIMUM ITERATION LIMIT...',I5,2X,                   INTL1500
    &', OUTPUT PRINTED EVERY',I4,' ITERATION(S)')                       INTL1510
1060 FORMAT(/,20X,I5,' IS THE MAXIMUM NUMBER OF UNIDIRECTIONAL SEARCH'  INTL1520
    &,'ES DURING EACH',/,20X,'MAXIMIZATION PHASE (MULTIPLIERS AND '     INTL1530
    &,'WEIGHTS ARE HELD CONSTANT WITHIN',/,20X,'EACH SUCH PHASE).')     INTL1540
1070 FORMAT(/,20X,'INITIAL PENALTY WEIGHTS ARE ',3(F7.2,2X),//,         INTL1550
    &20X,'RESPECTIVELY, AND ARE UPDATED BY A FACTOR OF ',F7.2,          INTL1560
    &' UNTIL THEY EXCEED',/,20X,'VALUES OF ',3(D10.3,2X))               INTL1570
1090 FORMAT(I5,D15.5,I5,D15.5,I5,D15.5,I5,D15.5)                       INTL1580
1100 FORMAT(I5,I5,D15.5,D15.5,I5,D15.5,I5,D15.5)                       INTL1590
1110 FORMAT(/,20X,'X VALUES, X(1),...,X(N)')                           INTL1600
1120 FORMAT(/,20X,'$$$ F = ',D13.6,3X,'AUG. LAG. = ',D13.6,3X,         INTL1610
    &'GMAG = ',D13.6,/,20X,'###############################           INTL1620
    &##############################')                                   INTL1630
1130 FORMAT(20X,D13.6,1X,D13.6,1X,D13.6,1X,D13.6,1X,D13.6)             INTL1640
1140 FORMAT(20X,5X,'UNCONSTRAINED')                                    INTL1650
1150 FORMAT(20X,I5,4(3X,D13.6))                                        INTL1660
1160 FORMAT('1')                                                        INTL1670
1170 FORMAT(18A4)                                                      INTL1680
1180 FORMAT(20X,18A4)                                                  INTL1690
1190 FORMAT(/,20X,'EQUALITY CONSTRAINT STATUS:')                       INTL1700
```

```
 1200 FORMAT(20X,2X,'I',9X,'FE(I)',12X,'A(I)',9X,'A(I)-FE(I)',          INTL1710
     &8X,'HE(I)')                                                        INTL1720
 1210 FORMAT(20X,'WEIGHT/MULTIPLIER UPDATE INFORMATION SUPPRESSED')      INTL1730
 1220 FORMAT(20X,'WEIGHT/MULTIPLIER UPDATE INFORMATION IS PRINTED')      INTL1740
 1230 FORMAT(/,20X,'INEQUALITY CONSTRAINT STATUS:')                      INTL1750
 1240 FORMAT(20X,2X,'J',9X,'FI(J)',12X,'B(J)',9X,'B(J)-FI(J)',           INTL1760
     &8X,'HI(J)')                                                        INTL1770
 1250 FORMAT(/,20X,'CONSTRAINED VARIABLE STATUS:')                       INTL1780
 1260 FORMAT(20X,4X,'HVL(K)',12X,'C(K)',9X,'K',10X,'D(K)',12X,           INTL1790
     &'HVU(K)')                                                          INTL1800
 1270 FORMAT(20X,D13.6,4X,D13.6,4X,I3)                                   INTL1810
 1280 FORMAT(20X,34X,I3,4X,D13.6,4X,D13.6)                               INTL1820
 1290 FORMAT(20X,D13.6,4X,D13.6,4X,I3,4X,D13.6,4X,D13.6)                 INTL1830
 1300 FORMAT(/,20X,3X,'NA',6X,'NB',6X,'IGF',5X,'IGE',5X,'IGI',5X,        INTL1840
     &'IHE',5X,'IHI',4X,'IHVL',4X,'IHVU',/,20X,9(I5,3X),/)               INTL1850
 1310 FORMAT(/,20X,'ISS = ',I1,10X,'IRESET = ',I1)                       INTL1860
      END                                                                INTL1870
```

B FORTRAN SUBROUTINES

```
      SUBROUTINE CONGR(GT,GO,IX,N,NT)
C
C     PRINTS CONSTANT GRADIENT COMPONENTS
C
      DIMENSION GT(1),GO(1),IX(1)
C
      PRINT 1000
      DO 20 I=1,NT
      INDX=0
      I1=(I-1)*N + 1
      I2=I*N
      DO 10 K=I1,I2
      IF(GT(K).EQ.0.) GO TO 10
      INDX=INDX+1
      IX(INDX)=K
      GO(INDX)=GT(K)
   10 CONTINUE
      IF(INDX.EQ.0) GO TO 20
      PRINT 1010,(IX(L),GO(L),L=1,INDX)
   20 CONTINUE
      RETURN
C
C     ********* FORMATS **********
C
 1000 FORMAT(/,20X,'CONSTANT GRADIENT COMPONENTS OF THE FUNCTION(S)')
 1010 FORMAT(20X,I4,D14.6,I4,D14.6,I4,D14.6,I4,D14.6)
      END
```

```
      SUBROUTINE DFFRV(GAL,H,DLX,DLG,N,R)                               DFFPR0010
C                                                                       DFFPR0020
C     COMPUTES SEARCH DIRECTION R, VIA DFP METHOD WITH RESETS OF        DFFPR0030
C     R = GAL BASED UPON UPDATE CRITERION, ETC.                         DFFPR0040
C                                                                       DFFPR0050
      DIMENSION GAL(1),DLX(1),DLG(1),R(1),H(1)                          DFFPR0060
      COMMON /C9/KSRCH,IBSD/C13/RMAG,RHO,GMAG,SLOPE                     DFFPR0070
      COMMON /C18/NG,IUP,IFAIL/C25/ISS,IRESET,NN                        DFFPR0080
      NG=NG+1                                                           DFFPR0090
      IF(RHO.EQ.0.) GO TO 10                                            DFFPR0100
      IF(KSRCH.EQ.0 .AND. IFAIL.GT.0) GO TO 10                          DFFPR0110
      IF(IRESET.EQ.0) GO TO 40                                          DFFPR0120
      IF(KSRCH.NE.0 .OR. NG.LE.N) GO TO 40                              DFFPR0130
   10 NG=1                                                              DFFPR0140
                                                                        DFFPR0150
      R = GAL                                                           DFFPR0160
C                                                                       DFFPR0170
C                                                                       DFFPR0180
      SLOPE=0.                                                          DFFPR0190
      DO 20 I=1,NN                                                      DFFPR0200
   20 H(I)=0.                                                           DFFPR0210
      INDX=0                                                            DFFPR0220
      DO 30 I=1,N                                                       DFFPR0230
      H(INDX+I)=1.                                                      DFFPR0240
      R(I)=GAL(I)                                                       DFFPR0250
   30 SLOPE=SLOPE+R(I)**2                                               DFFPR0260
      INDX=INDX+N                                                       DFFPR0270
      RMAG=SQRT(SLOPE)                                                  DFFPR0280
      RETURN                                                            DFFPR0290
```

```
C     R VIA DFF OR DFFSS
C
C
   40 CON1=0.                                                           DFFPR0300
      CON2=0.                                                           DFFPR0310
      CON3=0.                                                           DFFPR0320
      CON4=0.                                                           DFFPR0330
      INDX=0                                                            DFFPR0340
      DO 50 I=1,N                                                       DFFPR0350
      TFM=0.                                                            DFFPR0360
      DO 45 J=1,N                                                       DFFPR0370
   45 TFM=TFM + H(INDX+J)*DLG(J)                                        DFFPR0380
      R(I)=TFM
      CON1=CON1 + DLX(I)*DLG(I)                                         DFFPR0390
      CON2=CON2 + DLG(I)*R(I)                                           DFFPR0400
      IF(ISS.EQ.0) GO TO 50                                             DFFPR0410
      TEMP=GAL(I)-DLG(I)                                                DFFPR0420
      CON3=CON3 + TEMP*DLX(I)                                           DFFPR0430
      CON4=CON4 + TEMP*R(I)                                             DFFPR0440
   50 INDX=INDX+N                                                       DFFPR0450
      IF(CON1.EQ.0. .OR. CON2.EQ.0.) GO TO 10                           DFFPR0460
      IF(ISS.EQ.0. .OR. CON4.EQ.0.) GO TO 56                            DFFPR0470
      DO 55 I=1,N                                                       DFFPR0480
   55 DLG(I) = R(I)/CON2 - DLX(I)/CON1                                  DFFPR0490
      CON5 = .5*CON2                                                    DFFPR0500
      GAMA = -CON3/CON4                                                 DFFPR0510
   56 INDX=0                                                            DFFPR0520
      DO 70 K=1,N                                                       DFFPR0530
      ITRN=(K-1)*N                                                      DFFPR0540
                                                                        DFFPR0550
                                                                        DFFPR0560
                                                                        DFFPR0570
```

```
      DO 60 J=K,N
      L=INDX+J
      IF(ISS.EQ.0 .OR. CON4.EQ.0.) GO TO 57
      H(L) = ( H(L) - R(K)*R(J)/CON2 + CONS*DLG(K)*DLG(J) )*GAMA
     & - DLX(K)*DLX(J)/CON1
      GO TO 58
   57 H(L) = H(L) - DLX(K)*DLX(J)/CON1 - R(K)*R(J)/CON2
   58 H(ITRN+K) = H(L)
   60 ITRN=ITRN+N
   70 INDX=INDX+N
      INDX=0
      DO 80 I=1,N
      TFM=0.
      DO 75 J=1,N
   75 TFM=TFM + H(INDX+J)*GAL(J)
      R(I)=TFM
   80 INDX=INDX+N
C
C     TEST FOR A GOOD R; IF R BAD, SET R = GAL
C
      RMAG=0.
      SLOPE=0.
      DO 100 I=1,N
      RMAG=RMAG+R(I)*R(I)
  100 SLOPE=SLOPE+R(I)*GAL(I)
      RMAG=SQRT(RMAG)
      IF (SLOPE.GT.0.) RETURN
      GO TO 10
      END
```

B   FORTRAN SUBROUTINES   343

```
      SUBROUTINE SEARCH(A,B,C,D,FE,FES,FI,FIS,HE,HI,HVL,HVU,IX,     SRCH0010
     &X,XT,R)                                                       SRCH0020
C                                                                   SRCH0030
C     CONDUCTS SEARCH ALONG R TO FIND STEPSIZE RHO THAT MAXIMIZES   SRCH0040
C     THE AUGMENTED LAGRANGIAN. AL = F(X + RHO*R)                   SRCH0050
C                                                                   SRCH0060
      DIMENSION A(1),B(1),C(1),D(1),FE(1),FES(1),FI(1),FIS(1)       SRCH0070
      DIMENSION HE(1),HI(1),HVL(1),HVU(1),IX(1),X(1),XT(1),R(1)     SRCH0080
      COMMON /C1/N,NE,NI,NBV,NEPI/C2/NFE,NGE/C3/W1,W2,W3            SRCH0090
      COMMON /C13/RMAG,RHO,GMAG,SLOPE/C18/NG,IUP,IFAIL              SRCH0100
      COMMON /C9/KSRCH,IESD/C17/DS,FS,YS/C21/F,AL                   SRCH0110
C                                                                   SRCH0120
C     SAVE STARTING POINT (D1,Y1), SET INITIAL STEP D2, EVALUATE Y2 SRCH0130
C                                                                   SRCH0140
      DS=0.                                                         SRCH0150
      YS=AL                                                         SRCH0160
      D1=0.                                                         SRCH0170
      IC1=0                                                         SRCH0180
      Y1=AL                                                         SRCH0190
      IUP=0                                                         SRCH0200
      D2=1.                                                         SRCH0210
      IF(NG.EQ.1) D2=.1                                             SRCH0220
      IF(KSRCH.EQ.0 .AND. IFAIL.GT.0) D2=0.05                       SRCH0230
      IF(RMAG.LT.200.) GO TO 10                                     SRCH0240
      D2=10./RMAG                                                   SRCH0250
      IF(D2.GT.(.001)) GO TO 10                                     SRCH0260
      D2=.001                                                       SRCH0270
```

```
   10 CALL VALUE(A,B,C,D,FE,FES,FI,FIS,HE,HI,HVL,HVU,IX,         SRCH0280
     &X,XT,R,F,D2,Y2)                                             SRCH0290
C                                                                 SRCH0300
C     UNIDIRECTIONAL SEARCH                                       SRCH0310
C                                                                 SRCH0320
      CALL DMAX(D1,D2,D3,Y1,Y2,Y3,SLOPE,1,D3)                     SRCH0330
      IF (Y2-Y1) 50,200,20                                        SRCH0340
   20 IF (D3.LE.0.) GO TO 200                                     SRCH0350
      IF (D3.GE.(.9*D2).AND.D3.LE.(1.1*D2)) GO TO 410             SRCH0360
   25 IF (D3.LE.(5.*D2)) GO TO 340                                SRCH0370
      IF (D3.GT.(100.*D2)) D3=100.*D2                             SRCH0380
      CALL VALUE(A,B,C,D,FE,FES,FI,FIS,HE,HI,HVL,HVU,IX,          SRCH0390
     &X,XT,R,F,D3,Y3)                                             SRCH0400
      IF (Y3-Y2) 200,200,30                                       SRCH0410
   30 D2=D3                                                       SRCH0420
      Y2=Y3                                                       SRCH0430
      GO TO 300                                                   SRCH0440
   50 IF (D3.GT.(0.2*D2)) GO TO 100                               SRCH0450
      IF (D3.LT.(0.01*D2)) D3=0.01*D2                             SRCH0460
      CALL VALUE(A,B,C,D,FE,FES,FI,FIS,HE,HI,HVL,HVU,IX,          SRCH0470
     &X,XT,R,F,D3,Y3)                                             SRCH0480
      D2=D3                                                       SRCH0490
      Y2=Y3                                                       SRCH0500
      IF (Y2-Y1) 60,200,300                                       SRCH0510
   40 IF (IC1.GE.10) GO TO 410                                    SRCH0520
      IC1=IC1+1                                                   SRCH0530
      CALL DMAX(D1,D2,D3,Y1,Y2,Y3,SLOPE,1,D3)                     SRCH0540
      GO TO 50                                                    SRCH0550
```

B FORTRAN SUBROUTINES

```fortran
100 CALL VALUE(A,B,C,D,FE,FES,FI,FIS,HE,HI,HVL,HVU,IX,
   &X,XT,R,F,D3,Y3)
    IF (Y3-Y1) 120,410,350
120 D2=0.2*D3
    IC1=IC1+1
    CALL VALUE(A,B,C,D,FE,FES,FI,FIS,HE,HI,HVL,HVU,IX,
   &X,XT,R,F,D2,Y2)
    IF (Y2-Y1) 150,410,240
150 IF (IC1.GT.10) GO TO 240
160 D3=D2
    Y3=Y2
    GO TO 120
200 D3=5.*D2
    IC1=IC1+1
    CALL VALUE(A,B,C,D,FE,FES,FI,FIS,HE,HI,HVL,HVU,IX,
   &X,XT,R,F,D3,Y3)
    IF (Y3-Y2) 220,210,210
210 IF (IC1.GT.10) GO TO 240
    D1=D2
    Y1=Y2
    D2=D3
    Y2=Y3
    GO TO 200
220 IF (IUF) 410,410,240
240 CALL DMAX(D1,D2,D3,Y1,Y2,Y3,SLOPE,2,RHO)
    GO TO 400
300 CALL DMAX(D1,D2,D3,Y1,Y2,Y3,SLOPE,1,D3)
```

```
      IF (D3.GT.(0.9*D2).AND.D3.LT.(1.1*D2)) GO TO 410        SRCH0830
      IF (D3.LE.0.) GO TO 200                                 SRCH0840
      IF (IC1.GE.10) GO TO 410                                SRCH0850
      IC1=IC1+1                                               SRCH0860
      GO TO 25                                                SRCH0870
  340 CALL VALUE(A,B,C,D,FE,FES,FI,FIS,HE,HI,HVL,HVU,IX,      SRCH0880
     &X,XT,R,F,D3,Y3)                                         SRCH0890
      IF (Y3-Y2) 240,350,350                                  SRCH0900
  350 CALL DMAX(D1,D2,D3,Y1,Y2,Y3,SLOPE,2,RHO)                SRCH0910
      IF (RHO.GT.(.9*D3).AND.RHO.LT.(1.1*D3)) GO TO 410       SRCH0920
                                                              SRCH0930
C                                                             SRCH0940
C     BEST STEPSIZE RHO FOUND FOR THIS SEARCH                 SRCH0950
C                                                             SRCH0960
  400 CALL VALUE(A,B,C,D,FE,FES,FI,FIS,HE,HI,HVL,HVU,IX,      SRCH0970
     &X,XT,R,F,RHO,AL)                                        SRCH0980
  410 RHO=DS                                                  SRCH0990
      DO 420 I=1,N                                            SRCH1000
  420 XT(I)=X(I) + RHO*R(I)                                   SRCH1010
      IF(NE.EQ.0) GO TO 440                                   SRCH1020
      DO 430 I=1,NE                                           SRCH1030
  430 FE(I) = FES(I)                                          SRCH1040
  440 IF(NI.EQ.0) GO TO 460                                   SRCH1050
      DO 450 I=1,NI                                           SRCH1060
  450 FI(I) = FIS(I)                                          SRCH1070
  460 F=FS                                                    SRCH1080
      AL=YS                                                   SRCH1090
      RETURN                                                  SRCH1100
      END
```

B FORTRAN SUBROUTINES

```fortran
      SUBROUTINE VALUE(A,B,C,D,FE,FES,FI,FIS,HE,HI,HVL,HVU,IX,       VALU0010
     &X,XT,R,F,DT,Y)                                                 VALU0020
C                                                                    VALU0030
C     EVALUATES AUGMENTED LAGRANGIAN Y, AT STEPSIZE D,               VALU0040
C     ALONG SEARCH DIRECTION R; SAVES POINT IF BETTER                VALU0050
C                                                                    VALU0060
      DIMENSION A(1),B(1),C(1),D(1),FE(1),FES(1),FI(1),FIS(1)        VALU0070
      DIMENSION HE(1),HI(1),HVL(1),HVU(1),IX(1),X(1),XT(1),R(1)      VALU0080
      COMMON /C1/N,NE,NI,NBV,NEPI/C2/NFE,NGE/C3/W1,W2,W3             VALU0090
      COMMON /C17/DS,FS,YS/C18/NG,IUP,IFAIL                          VALU0100
C                                                                    VALU0110
      DO 5 I=1,N                                                     VALU0120
    5 XT(I) = X(I) + DT*R(I)                                         VALU0130
      CALL FXNS(XT,F,FE,FI)                                          VALU0140
      NFE=NFE+1                                                      VALU0150
      CALL AUGLAG(A,B,C,D,FE,FI,HE,HI,HVL,HVU,IX,XT,F,Y)             VALU0160
      IF(Y-YS) 60,15,10                                              VALU0170
   10 IUP=1                                                          VALU0180
   15 IF(NE.EQ.0) GO TO 30                                           VALU0190
      DO 20 I=1,NE                                                   VALU0200
   20 FES(I)=FE(I)                                                   VALU0210
   30 IF(NI.EQ.0) GO TO 50                                           VALU0220
      DO 40 I=1,NI                                                   VALU0230
   40 FIS(I)=FI(I)                                                   VALU0240
   50 FS=F                                                           VALU0250
      DS=DT                                                          VALU0260
      YS=Y                                                           VALU0270
   60 RETURN                                                         VALU0280
      END                                                            VALU0290
```

```fortran
      SUBROUTINE DMAX(D1,D2,D3,Y1,Y2,Y3,SLOPE,K,ALF)
C
C     ASSUMES QUADRATIC FORM   AL = A*(D1-ALF)**2 + B*(D1-ALF) + C
C     MAXIMUM ALF = D1 - B/(2*A) FOUND BY QUADRATIC FIT OF
C     DATA:  (POINT/SLOPE,POINT) OR (3 POINTS)
C
      GO TO (10,20),K
C
C     QUADRATIC FIT, PSP
C
   10 TP1=-SLOPE*D2
      DIF=Y2-Y1+TP1
      IF(DIF.EQ.0.) GO TO 30
      ALF=(.5*TP1*D2)/DIF
      RETURN
C
C     QUADRATIC FIT, 3P
C
   20 D21=D2-D1
      D31=D3-D1
      TP1=D31*(Y2-Y1)
      TP2=D21*(Y3-Y1)
      DIF=TP1-TP2
      IF(DIF.EQ.0.) GO TO 30
      ALF=D1-.5*(D21*TP2-D31*TP1)/DIF
      RETURN
C
C     ZERO DIVISOR
C
   30 ALF=25.*D2
      RETURN
      END
```

B FORTRAN SUBROUTINES

```
      SUBROUTINE DELTA(XN, XO, GN, GO, DLX, DLG, N)                     DLTA0010
                                                                        DLTA0020
C     COMPUTES:  NORM(XNEW-XOLD), NORM(GNEW-GOLD), NORM(GNEW)           DLTA0030
                                                                        DLTA0040
      DIMENSION XN(1), XO(1), GN(1), GO(1), DLX(1), DLG(1)              DLTA0050
      COMMON /C6/DELX, DELG/C13/RMAG, RHO, GMAG, SLOPE                  DLTA0060
                                                                        DLTA0070
      DELG=0.                                                           DLTA0080
      GMAG=0.                                                           DLTA0090
      DO 10 I=1,N                                                       DLTA0100
      DLX(I)=XN(I)-XO(I)                                                DLTA0110
      DLG(I)=GN(I)-GO(I)                                                DLTA0120
      DELG=DELG+DLG(I)*DLG(I)                                           DLTA0130
      GMAG=GMAG+GN(I)*GN(I)                                             DLTA0140
      XO(I)=XN(I)                                                       DLTA0150
   10 GO(I)=GN(I)                                                       DLTA0160
      DELX=RHO*RMAG                                                     DLTA0170
      DELG=SQRT(DELG)                                                   DLTA0180
      GMAG=SQRT(GMAG)                                                   DLTA0190
      RETURN                                                            DLTA0200
      END                                                               DLTA0210
```

```
      SUBROUTINE UPDATE(A,B,C,D,FE,FI,HE,HI,HVL,HVU,IX,X)            UPDT0010
C                                                                   UPDT0020
C     MULTIPLIERS AND PENALTY WEIGHTS UPDATED BY UPDATE RULES        UPDT0030
C                                                                   UPDT0040
      DIMENSION A(1),B(1),C(1),D(1),FE(1),FI(1),HE(1),HI(1)          UPDT0050
      DIMENSION HVL(1),HVU(1),IX(1),X(1)                             UPDT0060
      COMMON /C1/N,NE,NI,NBV,NEPI/C3/W1,W2,W3                        UPDT0070
      COMMON /C4/W1MAX,W2MAX,W3MAX,WF/C20/NUF                        UPDT0080
C                                                                   UPDT0090
      TW1=W1+W1                                                      UPDT0100
      TW2=W2+W2                                                      UPDT0110
      TW3=W3+W3                                                      UPDT0120
C                                                                   UPDT0130
C     EQUALITY CONSTRAINT MULTIPLIERS                                UPDT0140
C                                                                   UPDT0150
      IF(NE.EQ.0) GO TO 20                                           UPDT0160
      DO 10 I=1,NE                                                   UPDT0170
   10 HE(I)=HE(I)-TW1*(A(I)-FE(I))                                   UPDT0180
C                                                                   UPDT0190
C     INEQUALITY CONSTRAINT MULTIPLIERS                              UPDT0200
C                                                                   UPDT0210
   20 IF (NI.EQ.0) GO TO 50                                          UPDT0220
      DO 40 J=1,NI                                                   UPDT0230
      TEMP=B(J)-FI(J)                                                UPDT0240
      IF(HI(J).LE.0.) GO TO 30                                       UPDT0250
      HI(J)=HI(J)-TW2*TEMP                                           UPDT0260
      IF(HI(J).LT.0.) HI(J)=0.                                       UPDT0270
      GO TO 40                                                       UPDT0280
   30 IF(TEMP.LT.0.) HI(J)=-TW3*TEMP                                 UPDT0290
   40 CONTINUE                                                       UPDT0300
```

```
C     BOUNDED VARIABLES                                              UPDT0310
C                                                                    UPDT0320
C                                                                    UPDT0330
   50 IF(NBV.EQ.0) GO TO 200                                         UPDT0340
      DO 110 K=1,N                                                   UPDT0350
      IVAR=IX(K)                                                     UPDT0360
      IF(IVAR.EQ.0) GO TO 110                                        UPDT0370
      IF(IVAR.EQ.1.OR.IVAR.EQ.3) GO TO 70                            UPDT0380
   60 IF(IVAR.EQ.2.OR.IVAR.EQ.3) GO TO 90                            UPDT0390
      GO TO 110                                                      UPDT0400
C                                                                    UPDT0410
C     LOWER BOUND MULTIPLIERS                                        UPDT0420
C                                                                    UPDT0430
   70 TEMP=X(K)-C(K)                                                 UPDT0440
      IF(HVL(K).LE.0.) GO TO 80                                      UPDT0450
      HVL(K)=HVL(K)-TW2*TEMP                                         UPDT0460
      IF(HVL(K).LT.0.) HVL(K)=0.                                     UPDT0470
      GO TO 60                                                       UPDT0480
   80 IF(TEMP.LT.0.) HVL(K)=-TW3*TEMP                                UPDT0490
      GO TO 60                                                       UPDT0500
C                                                                    UPDT0510
C     UPPER BOUND MULTIPLIERS                                        UPDT0520
C                                                                    UPDT0530
   90 TEMP=D(K)-X(K)                                                 UPDT0540
      IF(HVU(K).LE.0.) GO TO 100                                     UPDT0550
      HVU(K)=HVU(K)-TW2*TEMP                                         UPDT0560
      IF(HVU(K).LT.0.) HVU(K)=0.                                     UPDT0570
      GO TO 110                                                      UPDT0580
```

```
      100 IF(TEMP.LT.0.) HVU(K)=-TW3*TEMP              UPDT0590
      110 CONTINUE                                     UPDT0600
    C                                                  UPDT0610
    C         PENALTY WEIGHT UPDATE                    UPDT0620
    C                                                  UPDT0630
      200 IF(NE.EQ.0) GO TO 210                        UPDT0640
          IF(W1.LT.W1MAX) W1=WF*W1                     UPDT0650
          IF(W1.GT.W1MAX) W1=W1MAX                     UPDT0660
      210 IF((NI+NEV).EQ.0) GO TO 300                  UPDT0670
          IF(W2.LT.W2MAX) W2=WF*W2                     UPDT0680
          IF(W2.GT.W2MAX) W2=W2MAX                     UPDT0690
          IF(W3.LT.W3MAX) W3=WF*W3                     UPDT0700
          IF(W3.GT.W3MAX) W3=W3MAX                     UPDT0710
    C                                                  UPDT0720
    C         PRINT UPDATE STATUS IF NUP=0             UPDT0730
    C                                                  UPDT0740
      300 IF(NUP.EQ.1) RETURN                          UPDT0750
          IF(NE.EQ.0) GO TO 310                        UPDT0760
          PRINT 1000                                   UPDT0770
          PRINT 1040,(FE(I),I=1,NE)                    UPDT0780
          PRINT 1010                                   UPDT0790
          PRINT 1040,(HE(I),I=1,NE)                    UPDT0800
      310 IF(NI.EQ.0) GO TO 320                        UPDT0810
          PRINT 1020                                   UPDT0820
          PRINT 1040,(FI(J),J=1,NI)                    UPDT0830
          PRINT 1030                                   UPDT0840
          PRINT 1040,(HI(J),J=1,NI)                    UPDT0850
      320 IF(NEV.EQ.0) GO TO 370                       UPDT0860
          PRINT 1050                                   UPDT0870
                                                       UPDT0880
```

B FORTRAN SUBROUTINES

```fortran
      DO 360 K=1,N
      IVAR=IX(K)+1                                                      UPDT0890
      GO TO (360,330,340,350),IVAR                                      UPDT0900
  330 VL=X(K)-C(K)                                                      UPDT0910
      PRINT 1060,HVL(K),VL,K                                            UPDT0920
      GO TO 360                                                         UPDT0930
  340 VU=D(K)-X(K)                                                      UPDT0940
      PRINT 1070,K,VU,HVU(K)                                            UPDT0950
      GO TO 360                                                         UPDT0960
  350 VL=X(K)-C(K)                                                      UPDT0970
      VU=D(K)-X(K)                                                      UPDT0980
      PRINT 1080,HVL(K),VL,K,VU,HVU(K)                                  UPDT0990
  360 CONTINUE                                                          UPDT1000
  370 PRINT 1090,W1,W2,W3                                               UPDT1010
      RETURN                                                            UPDT1020
C                                                                       UPDT1030
C     ********** FORMATS **********                                     UPDT1040
C                                                                       UPDT1050
 1000 FORMAT(/,20X,'EQUALITY CONSTRAINT VALUES, FE(1),...,FE(NE)')      UPDT1060
 1010 FORMAT(/,20X,'EQUALITY CONSTRAINT MULTIPLIERS, HE(1),...,HE(NE)') UPDT1070
 1020 FORMAT(/,20X,'INEQUALITY CONSTRAINT VALUES, FI(1),...,FI(NI)')    UPDT1080
 1030 FORMAT(/,20X,'INEQUALITY CONSTRAINT MULTIPLIERS,HI(1),...,HI(NI)')UPDT1090
 1040 FORMAT(20X,D13.6,1X,D13.6,1X,D13.6,1X,D13.6)                      UPDT1100
 1050 FORMAT(/,20X,13X,'LOWER',7X,'BOUNDED X(K) VARIABLES',7X,'UPPER',  UPDT1110
     &/,20X,4X,'HVL(K)',9X,'X(K)-C(K)',7X,'K',7X,'D(K)-X(K)',           UPDT1120
     &10X,'HVU(K)')                                                     UPDT1130
 1060 FORMAT(20X,D13.6,4X,D13.6,4X,I3)                                  UPDT1140
 1070 FORMAT(20X,34X,I3,4X,D13.6,4X,D13.6)                              UPDT1150
 1080 FORMAT(20X,D13.6,4X,D13.6,4X,I3,4X,D13.6,4X,D13.6)                UPDT1160
 1090 FORMAT(/,20X,'W1 = ',D13.6,5X,'W2 = ',D13.6,5X,'W3 = ',D13.6)     UPDT1170
      END                                                               UPDT1180
```

```
      SUBROUTINE AUGLAG(A,B,C,D,FE,FI,HE,HI,HVL,HVU,IX,X,F,AL)          ALAG0010
C                                                                       ALAG0020
C     FORMULATES AUGMENTED LAGRANGIAN - AL                              ALAG0030
C                                                                       ALAG0040
      DIMENSION A(1),B(1),C(1),D(1),FE(1),FI(1),HE(1),HI(1)             ALAG0050
      DIMENSION HVL(1),HVU(1),IX(1),X(1)                                ALAG0060
      COMMON /C1/N,NE,NI,NBV,NEFI/C3/W1,W2,W3                           ALAG0070
      IF(NEFI.EQ.0) GO TO 200                                           ALAG0080
      FLAG=0.                                                           ALAG0090
      FEW1=0.                                                           ALAG0100
      FIW2=0.                                                           ALAG0110
      FIW3=0.                                                           ALAG0120
C                                                                       ALAG0130
C     EQUALITY CONSTRAINT COMPONENTS                                    ALAG0140
C                                                                       ALAG0150
      IF(NE.EQ.0) GO TO 20                                              ALAG0160
      DO 10 I=1,NE                                                      ALAG0170
      TEMP=A(I)-FE(I)                                                   ALAG0180
      FLAG=FLAG+HE(I)*TEMP                                              ALAG0190
   10 FEW1=FEW1+TEMP*TEMP                                               ALAG0200
C                                                                       ALAG0210
C     INEQUALITY CONSTRAINT COMPONENTS                                  ALAG0220
C                                                                       ALAG0230
   20 IF(NI.EQ.0) GO TO 50                                              ALAG0240
      DO 40 J=1,NI                                                      ALAG0250
      TEMP=B(J)-FI(J)                                                   ALAG0260
      IF(HI(J).LE.0.) GO TO 30                                          ALAG0270
      FLAG=FLAG+HI(J)*TEMP                                              ALAG0280
      FIW2=FIW2+TEMP*TEMP                                               ALAG0290
                                                                        ALAG0300
```

```
      GO TO 40                                                          ALAG0310
   30 IF(TEMP.LT.0.) FIW3=FIW3+TEMP*TEMP                                ALAG0320
   40 CONTINUE                                                          ALAG0330
C                                                                       ALAG0340
C     BOUNDED VARIABLES                                                 ALAG0350
C                                                                       ALAG0360
   50 IF(NBV.EQ.0) GO TO 120                                            ALAG0370
      DO 110 K=1,N                                                      ALAG0380
      IVAR=IX(K)                                                        ALAG0390
      IF(IVAR.EQ.0) GO TO 110                                           ALAG0400
      IF(IVAR.EQ.1.OR.IVAR.EQ.3) GO TO 70                               ALAG0410
      IF(IVAR.EQ.2.OR.IVAR.EQ.3) GO TO 90                               ALAG0420
      GO TO 110                                                         ALAG0430
C                                                                       ALAG0440
C     LOWER BOUND COMPONENTS                                            ALAG0450
C                                                                       ALAG0460
   70 TEMP=X(K)-C(K)                                                    ALAG0470
      IF(HVL(K).LE.0.) GO TO 80                                         ALAG0480
      FLAG=FLAG+HVL(K)*TEMP                                             ALAG0490
      FIW2=FIW2+TEMP*TEMP                                               ALAG0500
      GO TO 60                                                          ALAG0510
   80 IF(TEMP.LT.0.) FIW3=FIW3+TEMP*TEMP                                ALAG0520
      GO TO 60                                                          ALAG0530
C                                                                       ALAG0540
C     UPPER BOUND COMPONENTS                                            ALAG0550
C                                                                       ALAG0560
   90 TEMP=D(K)-X(K)                                                    ALAG0570
      IF(HVU(K).LE.0.) GO TO 100                                        ALAG0580
      FLAG=FLAG+HVU(K)*TEMP                                             ALAG0590
```

```
      FIW2=FIW2+TEMP*TEMP                                               ALAG0600
      GO TO 110                                                         ALAG0610
  100 IF(TEMP.LT.0.) FIW3=FIW3+TEMP*TEMP                                ALAG0620
  110 CONTINUE                                                          ALAG0630
C                                                                       ALAG0640
C     AL EQUALS THE SUM OF THE COMPONENTS                               ALAG0650
C                                                                       ALAG0660
  120 AL = F + FLAG - W1*FEW1 - W2*FIW2 - W3*FIW3                       ALAG0670
      RETURN                                                            ALAG0680
  200 AL = F                                                            ALAG0690
      RETURN                                                            ALAG0700
      END                                                               ALAG0710

      SUBROUTINE GALAG(A,B,C,D,FE,FI,GAL,GE,GF,GI,HE,HI,HVL,HVU,IX,X)   GLAG0010
C                                                                       GLAG0020
C     FORMULATES THE GRADIENT OF THE AUGMENTED LAGRANGIAN - GAL         GLAG0030
C                                                                       GLAG0040
      DIMENSION A(1),B(1),C(1),D(1),FE(1),FI(1),GAL(1),GE(1)            GLAG0050
      DIMENSION GF(1),GI(1),HE(1),HI(1),HVL(1),HVU(1),IX(1),X(1)        GLAG0060
      COMMON /C1/N,NE,NI,NBV,NEFI/C3/W1,W2,W3                           GLAG0070
C                                                                       GLAG0080
      IF(NEFI.EQ.0) GO TO 200                                           GLAG0090
      DO 110 K=1,N                                                      GLAG0100
      GLAG=0.                                                           GLAG0110
      GEW1=0.                                                           GLAG0120
      GIW2=0.                                                           GLAG0130
      GIW3=0.                                                           GLAG0140
```

B   FORTRAN SUBROUTINES   357

```
C     EQUALITY CONSTRAINT COMPONENTS                           GLAG0150
C                                                              GLAG0160
C                                                              GLAG0170
      IF(NE.EQ.0) GO TO 20                                     GLAG0180
      INDX=K                                                   GLAG0190
      DO 10 I=1,NE                                             GLAG0200
      GLAG=GLAG+HE(I)*GE(INDX)                                 GLAG0210
      GEW1=GEW1+(A(I)-FE(I))*GE(INDX)                          GLAG0220
   10 INDX=INDX+N                                              GLAG0230
C                                                              GLAG0240
C     INEQUALITY CONSTRAINT COMPONENTS                         GLAG0250
C                                                              GLAG0260
   20 IF(NI.EQ.0) GO TO 50                                     GLAG0270
      INDX=K                                                   GLAG0280
      DO 40 J=1,NI                                             GLAG0290
      TEMP=B(J)-FI(J)                                          GLAG0300
      IF(HI(J).LE.0.) GO TO 30                                 GLAG0310
      GLAG=GLAG+HI(J)*GI(INDX)                                 GLAG0320
      GIW2=GIW2+TEMP*GI(INDX)                                  GLAG0330
      GO TO 40                                                 GLAG0340
   30 IF(TEMP.LT.0.) GIW3=GIW3+TEMP*GI(INDX)                   GLAG0350
   40 INDX=INDX+N                                              GLAG0360
C                                                              GLAG0370
C     BOUNDED VARIABLES                                        GLAG0380
C                                                              GLAG0390
   50 IF(NBV.EQ.0) GO TO 110                                   GLAG0400
      IVAR=IX(K)                                               GLAG0410
      IF(IVAR.EQ.0) GO TO 110                                  GLAG0420
      IF(IVAR.EQ.1.OR.IVAR.EQ.3) GO TO 70                      GLAG0430
   60 IF(IVAR.EQ.2.OR.IVAR.EQ.3) GO TO 90                      GLAG0440
      GO TO 110                                                GLAG0450
```

```
C     LOWER BOUND COMPONENTS                                            GLAG0460
C                                                                       GLAG0470
C                                                                       GLAG0480
   70 TEMP=X(K)-C(K)                                                    GLAG0490
      IF(HVL(K).LE.0.) GO TO 80                                         GLAG0500
      GLAG=GLAG-HVL(K)                                                  GLAG0510
      GIW2=GIW2-TEMP                                                    GLAG0520
      GO TO 60                                                          GLAG0530
   80 IF(TEMP.LT.0.) GIW3=GIW3-TEMP                                     GLAG0540
      GO TO 60                                                          GLAG0550
C                                                                       GLAG0560
C     UPPER BOUND COMPONENTS                                            GLAG0570
C                                                                       GLAG0580
   90 TEMP=D(K)-X(K)                                                    GLAG0590
      IF(HVU(K).LE.0.) GO TO 100                                        GLAG0600
      GLAG=GLAG+HVU(K)                                                  GLAG0610
      GIW2=GIW2+TEMP                                                    GLAG0620
      GO TO 110                                                         GLAG0630
  100 IF(TEMP.LT.0.) GIW3=GIW3+TEMP                                     GLAG0640
C                                                                       GLAG0650
C     GAL EQUALS THE SUM OF THE COMPONENTS                              GLAG0660
C                                                                       GLAG0670
  110 GAL(K) = GF(K) - GLAG + 2.*(W1*GEW1 + W2*GIW2 + W3*GIW3)          GLAG0680
      RETURN                                                            GLAG0690
  200 DO 210 K=1,N                                                      GLAG0700
  210 GAL(K) = GF(K)                                                    GLAG0710
      RETURN                                                            GLAG0720
      END                                                               GLAG0730
```

B FORTRAN SUBROUTINES

```
      SUBROUTINE OUTPUT(A,B,C,D,FE,FI,HE,HI,HVL,HVU,IX,X,R)            OUTP0010
C                                                                      OUTP0020
C     OUTPUTS GENERAL AND TERMINAL SEARCH POINT INFORMATION            OUTP0030
C                                                                      OUTP0040
      DIMENSION A(1),B(1),C(1),D(1),FE(1),FI(1),HE(1),HI(1)            OUTP0050
      DIMENSION HVL(1),HVU(1),IX(1),X(1),R(1)                          OUTP0060
      COMMON /C1/N,NE,NI,NBV,NEPI/C2/NFE,NGE/C3/W1,W2,W3               OUTP0070
      COMMON /C6/DELX,DELG/C11/ISTOP,ITOT/C21/F,AL                     OUTP0080
      COMMON /C13/RMAG,RHO,GMAG,SLOPE/C18/NG,IUP,IFAIL                 OUTP0090
C                                                                      OUTP0100
C     GENERAL SEARCH POINT INFORMATION                                 OUTP0110
C                                                                      OUTP0120
      PRINT 1000,ITOT                                                  OUTP0130
      PRINT 1010,NFE,NGE,RHO                                           OUTP0140
      PRINT 1050,F,AL,GMAG                                             OUTP0150
      IF(ISTOP.EQ.0.AND.RHO.EQ.0.) GO TO 110                           OUTP0160
      PRINT 1030,NG                                                    OUTP0170
      PRINT 1040,(R(I),I=1,N)                                          OUTP0180
      PRINT 1020,DELX,DELG,RMAG                                        OUTP0190
      PRINT 1060                                                       OUTP0200
      PRINT 1040,(X(I),I=1,N)                                          OUTP0210
      IF(ISTOP.EQ.0) RETURN                                            OUTP0220
C                                                                      OUTP0230
C     TERMINAL INFORMATION                                             OUTP0240
C                                                                      OUTP0250
      IF(NEPI.EQ.0) GO TO 100                                          OUTP0260
      IF(NE.EQ.0) GO TO 20                                             OUTP0270
```

```
      PRINT 1190                                                    OUTF0280
      PRINT 1200                                                    OUTF0290
      DO 10 I=1,NE                                                  OUTF0300
      TEMP=A(I)-FE(I)                                               OUTF0310
   10 PRINT 1150,I,FE(I),A(I),TEMP,HE(I)                            OUTF0320
   20 IF(NI.EQ.0) GO TO 40                                          OUTF0330
      PRINT 1230                                                    OUTF0340
      PRINT 1240                                                    OUTF0350
      DO 30 J=1,NI                                                  OUTF0360
      TEMP=B(J)-FI(J)                                               OUTF0370
   30 PRINT 1150,J,FI(J),B(J),TEMP,HI(J)                            OUTF0380
   40 IF(NEV.EQ.0) GO TO 90                                         OUTF0390
      PRINT 1250                                                    OUTF0400
      PRINT 1100                                                    OUTF0410
      DO 80 K=1,N                                                   OUTF0420
      IVAR=IX(K)+1                                                  OUTF0430
      GO TO (80,50,60,70),IVAR                                      OUTF0440
   50 VL=X(K)-C(K)                                                  OUTF0450
      PRINT 1090,HVL(K),VL,K                                        OUTF0460
      GO TO 80                                                      OUTF0470
   60 VU=D(K)-X(K)                                                  OUTF0480
      PRINT 1070,K,VU,HVU(K)                                        OUTF0490
      GO TO 80                                                      OUTF0500
   70 VL=X(K)-C(K)                                                  OUTF0510
      VU=D(K)-X(K)                                                  OUTF0520
      PRINT 1080,HVL(K),VL,K,VU,HVU(K)                              OUTF0530
   80 CONTINUE                                                      OUTF0540
```

B FORTRAN SUBROUTINES

```
   90 PRINT 1130,W1,W2,W3                                              OUTP0550
  100 PRINT 1110                                                        OUTP0560
      PRINT 1120                                                        OUTP0570
      RETURN                                                            OUTP0580
  110 PRINT 1260                                                        OUTP0590
      RETURN                                                            OUTP0600
C                                                                       OUTP0610
C     **********     FORMATS     **********                             OUTP0620
C                                                                       OUTP0630
 1000 FORMAT(///,20X,'********************** SEARCH POINT ***',I6,      OUTP0640
     &' **********************')                                        OUTP0650
 1010 FORMAT(/,20X,'FXNS CALLS ... ',I6,5X,'GRAD CALLS ... ',I6,5X      OUTP0660
     &'RHO = ',D13.6)                                                   OUTP0670
 1020 FORMAT(/,20X,'DELX = ',D13.6,6X,'DELG = ',D13.6,6X,               OUTP0680
     &'RMAG = ',D13.6)                                                  OUTP0690
 1030 FORMAT(/,20X,'SEARCH DIRECTION TO THIS POINT, R(1),...,R(N)',     OUTP0700
     &10X,'NG =',I4)                                                    OUTP0710
 1040 FORMAT(20X,D13.6,1X,D13.6,1X,D13.6,1X,D13.6)                      OUTP0720
 1050 FORMAT(/,20X,'$$$ F = ',D13.6,3X,'AUG. LAG. = ',D13.6,3X,         OUTP0730
     &'GMAG = ',D13.6)                                                  OUTP0740
 1060 FORMAT(/,20X,'X VALUES, X(1),...,X(N)')                           OUTP0750
 1070 FORMAT(20X,34X,I3,4X,D13.6,4X,D13.6)                              OUTP0760
 1080 FORMAT(20X,D13.6,4X,D13.6,4X,I3,4X,D13.6,4X,D13.6)                OUTP0770
 1090 FORMAT(20X,D13.6,4X,D13.6,4X,I3)                                  OUTP0780
 1100 FORMAT(/,20X,13X,'LOWER',7X,'BOUNDED X(K) VARIABLES',7X,'UPPER',  OUTP0790
     &/,20X,4X,'HVL(K)',9X,'X(K)-C(K)',7X,'K',7X,'D(K)-X(K)',           OUTP0800
     &10X,'HVU(K)')                                                     OUTP0810
```

```
1110 FORMAT(/,20X,'*********************************************
     &*************')
1120 FORMAT('1')
1130 FORMAT(/,20X,'W1 = ',D13.6,5X,'W2 = ',D13.6,5X,'W3 = ',D13.6)
1150 FORMAT(20X,I5,4(3X,D13.6))
1190 FORMAT(/,20X,'EQUALITY CONSTRAINT STATUS:')
1200 FORMAT(20X,2X,'I',9X,'FE(I)',12X,'A(I)',9X,'A(I)-FE(I)',
     &8X,'HE(I)')
1230 FORMAT(/,20X,'INEQUALITY CONSTRAINT STATUS:')
1240 FORMAT(20X,2X,'J',9X,'FI(J)',12X,'B(J)',9X,'B(J)-FI(J)',
     &8X,'HI(J)')
1250 FORMAT(/,20X,'CONSTRAINED VARIABLE STATUS:')
1260 FORMAT(/,20X,'SEARCH FAILED TO FIND A LARGER VALUE OF ',
     &'THE AUGMENTED LAGRANGIAN')
     END
```

APPENDIX C

C MODIFIED LAGRANGIAN FORMS

Over two centuries have elapsed since Lagrange's fundamental work on the use of multipliers for treating equality constraints (Joseph L. Lagrange, 1736-1813). During that time, much has been done to generalize the multiplier approach. Several recent generalizations are associated with the modified Lagrangian forms that are listed in this appendix.

Consider the following problem:

maximize $f(x)$ (C-1)

subject to

    $p_i(x) = 0, \quad i = 1, 2, \ldots, m_1 < n,$ and   (C-2)

    $q_j(x) \leq 0, \quad j = 1, 2, \ldots, m_2$ (C-3)

where p_i's and q_j's are of class C^1. Note that

this is the same problem as that in Section 1.2, except that the right-hand sides of the constraints have been shifted to the left and incorporated in the constraint functions of (C-2) and (C-3).

The following definitions are in order:

$\phi = \phi(p_i, \alpha_i, w)$ is a multiplier function associated with the ith equality constraint function $p_i(x) = 0$.

$\lambda = \lambda(q_j, \beta_j, w)$ is a multiplier function associated with the jth inequality constraint function $q_j(x) \leq 0$.

α is an equality constraint multiplier vector of dimension m_1.

β is an equality constraint multiplier vector of dimension m_2.

w is a weighting parameter, $w > 0$.

θ_i and t are parameters.

The generalized modified Lagrangian, L_m, is also defined:

$$L_m(x, \alpha, \beta, w) \triangleq f(x) - \sum_{i=1}^{m_1} \phi(p_i, \alpha_i, w)$$

$$- \sum_{j=1}^{m_2} \lambda(q_j, \beta_j, w) \qquad (C-4)$$

C MODIFIED LAGRANGIAN FORMS

In the following, representative forms of ϕ and λ functions are listed. The notation used and the signs of some of the functions differ here from the original sources, in order to apply to the maximization problem of (C-1), (C-2), and (C-3). Examples of multiplier functions for the ϕ class are:

M1: $\alpha_i p_i + w p_i^2$ (Hestenes) (Arrow, Solow)

M2: $w(p_i + \theta_i)^2$ where $2w\theta_i = \alpha_i$ (Powell)

M3: $\dfrac{1}{tw}[(wp_i + \alpha_i)^t - \alpha_i^t]$ (Mangasarian)

    where t is an even integer, $t \geq 2$.

M4: $\cosh(wp_i + \alpha_i) - \tfrac{1}{2}(wp_i + \alpha_i)^2$

    $- \cosh(\alpha_i) + \tfrac{1}{2}\alpha_i^2$ (Mangasarian)

Examples of multiplier functions for the λ class are:

M5: $\beta_j [(1 + q_j)^t - 1]/t$ (Arrow, Hurwicz)

    where t is an odd integer and $t \geq 3$.

M6: $(\beta_j/w)[\exp(wq_j) - 1]$ (Gould, Howe)

M7: $\begin{cases} -\dfrac{\beta_j}{4w}, & q_j \leq -\dfrac{1}{2w} \\ \beta_j w q_j^2 + \beta_j q_j, & q_j > -\dfrac{1}{2w} \end{cases}$ (Arrow, Gould, Howe)

$$\text{M8:} \begin{cases} -\dfrac{\beta_j^{\,2}}{4w}, & q_j \le -\dfrac{\beta_j}{2w} \\ wq_j^{\,2} + \beta_j q_j, & q_j > -\dfrac{\beta_j}{2w} \end{cases} \qquad \text{(Rockafellar)}$$

$$\text{M9:} \begin{cases} \beta_j q_j + \tfrac{1}{2} wq_j^{\,2} + [(wq_j)^t/w], & 0 \le wq_j \\ \beta_j q_j + \tfrac{1}{2} wq_j^{\,2}, & -\beta_j < wq_j < 0 \\ -\tfrac{1}{2} wq_j^{\,2}, & wq_j \le -\beta_j \end{cases}$$

(Kort, Bertsekas)

where $t > 2$ selected arbitrarily.

In the following $z_+ = \begin{cases} z, & z \ge 0 \\ 0, & z < 0 \end{cases}$

M10: $\dfrac{1}{tw}\left((wq_j + \beta_j)_+^{\,t} - \beta_j^{\,t}\right)$

(Mangasarian)

where t is an even integer, $t \ge 4$.

M11: $\cosh(wq_j + \beta_j)_+ - \tfrac{1}{2}(wq_j + \beta_j)_+^{\,2}$
$\qquad - \cosh\beta_j + \tfrac{1}{2}\beta_j^{\,2}$

(Mangasarian)

M12: $\tfrac{1}{2}[\cosh(wq_j + \beta_j)_+ - 1]^2$

(Mangasarian)

$\qquad - \tfrac{1}{2}(\cosh\beta_j - 1)^2$

$$\text{M13:} \begin{cases} \beta_j q_j + wq_j^{\,2}, & q_j > 0 \\ \beta_j^{\,2} q_j/(\beta_j + wq_j), & q_j \le 0 \end{cases} \qquad \text{(Sayama et al.)}$$

C MODIFIED LAGRANGIAN FORMS

$$\text{M14:} \begin{cases} \beta_j q_j + w q_j^2, & \beta_j > 0 \\ w q_j (q_j + |q_j|)/2, & \beta_j = 0 \end{cases} \quad \text{(Pierre, Lowe)}$$

where β_j is updated such that $\beta_j \geq 0$.

Each of the functions above give the general form of their class of functions from which other multiplier functions may be generated. All of these functions have the property that $\nabla \phi$ and $\nabla \lambda$ are continuous functions of x. Function M13 has the property that $\nabla^2 \lambda$ is a continuous matrix function of x if q_j is of continuity class C^2. This desirable attribute is shared by M14 for the locked-on active constraints having $\beta_j > 0$.

APPENDIX D

D   MATRIX OPERATIONS

Selected properties of matrices and vectors are considered in this appendix. Detailed information on matrix theory can be found in various references ([B4, F1, L5, P2] for example).

MATRIX

A matrix is a rectangular array of entries which are usually numbers. Let Z be an m×n matrix given by

$$Z = \begin{bmatrix} z_{11} & z_{12} & \cdots & z_{1n} \\ \cdot & & & \cdot \\ \cdot & & & \cdot \\ \cdot & & & \cdot \\ z_{m1} & \cdot & \cdots & z_{mn} \end{bmatrix} \qquad (D-1)$$

D MATRIX OPERATIONS

which has m rows and n columns. An n×1 matrix is called a column vector, and a 1×n matrix is called a row vector. The matrix Z is also expressed as $[z_{ij}]$ where z_{ij} is the entry from the ith row in the jth column.

ADDITION

If Y and Z are matricies having the same dimensions,

$$Y + Z = Z + Y = [z_{ij} + y_{ij}] \qquad (D-2)$$

SCALAR MULTIPLICATION

Given a scalar γ,

$$\gamma Z = Z\gamma = [\gamma z_{ij}] \qquad (D-3)$$

MATRIX MULTIPLICATION

Given an m×n matrix Z and an n×ℓ matrix Y, the product ZY is dimensioned m×ℓ,

$$ZY = [\sum_{k=1}^{n} z_{ik} y_{kj}] \qquad (D-4)$$

IDENTITY MATRIX

An n×n matrix is an identity matrix I if its i,j^{th} entry is δ_{ij}, where

$$\delta_{ij} = \begin{cases} 1 & \text{for } i=j \\ 0 & \text{for } i \neq j \end{cases} \qquad (D-5)$$

Given any m×n matrix Z, ZI = Z.

TRANSPOSE

The transpose Z' of an m×n matrix Z is an n×m matrix, the i,j^{th} entry of which is z_{ji}. It follows that $(Z')' = Z$ and $(ZY)' = Y'Z'$.

SYMMETRIC MATRIX

Let Z be an n×n matrix. If $Z = Z'$, Z is a symmetric matrix.

DIAGONAL MATRIX

Let Z be an n×n matrix. If $z_{ij} = 0$ for all $i \neq j$, Z is a diagonal matrix and is represented by $Z = \text{diag}(z_{11}, z_{22}, \ldots, z_{nn})$.

D MATRIX OPERATIONS

PARTITIONED MATRIX

If one or more entries of a matrix Z is itself a matrix, Z is a partitioned matrix.

LINEAR COMBINATION

Let $r_1, r_2, \ldots, r_n$ be a set of vectors, and $\alpha_1, \alpha_2, \ldots, \alpha_r$ be a set of scalars. The vector $v = \sum_{i=1}^{n} \alpha_i r_i$ is a linear combination of the r_i's.

LINEAR INDEPENDENCE

A set of vectors is a linearly independent set if and only if the the only linear combination which is zero is the linear combination having all α_i's = 0.

RANK OF A MATRIX

The row rank of a matrix Z is the number of linearly independent rows; the column rank is the number of linearly independent columns. The row rank always equals the column rank, and is the rank of Z.

DETERMINANT OF A MATRIX

Let Z be an $n \times n$ matrix. The determinant $|Z|$ of Z is a scalar that can be computed as follows:

$$|Z| = \sum_{i=1}^{n} z_{ij} c_{ij}, \quad j \in \{1, 2, \ldots, n\} \qquad \text{(D-6)}$$

where

$$c_{ij} = (-1)^{i+j} |M_{ij}| \qquad \text{(D-7)}$$

and M_{ij} is the submatrix of Z obtained by deleting the ith row and j^{th} column of Z.

SINGULAR MATRIX

If $|Z| = 0$, the $n \times n$ matrix Z is singular. It can be shown that Z is singular if and only if the rank of Z is less than n.

QUADRATIC FORM

Let x be an $n \times 1$ vector of real variables, and let A be an $n \times n$ symmetric matrix of real numbers. The function $f(x) = x'Ax$ is a quadratic form:

$$x'Ax = \sum_{i=1}^{n} \sum_{j=1}^{n} a_{ij} x_i x_j \qquad \text{(D-8)}$$

D MATRIX OPERATIONS

POSITIVE DEFINITE (NEGATIVE DEFINITE)

An n×n matrix A of reals is said to be positive definite (negative definite) if $x'Ax > 0$ ($x'Ax < 0$) for all nontrivial n×1 x vectors. If $x'Ax \geq 0$ ($x'Ax \leq 0$) in the above statement, A is positive (negative) semidefinite.

SYLVESTER'S TEST

Given is an n×n symmetric matrix A. Let A_i denote the submatrix formed by deleting the last n-i rows and columns of A. Also, let $|A_i|$ denote the determinant of A_i, i = 1, 2, ..., n. Then

1) A is positive definite if $|A_i| > 0$ for i = 1, 2, ..., n.

2) A is negative definite if $|A_i| < 0$ for i = 1, 3, 5, ..., and $|A_i| > 0$ for i = 2, 4, 6, ...

If one or more $|A_i| = 0$, but the remaining inequalities of the above statement 1 (statement 2) apply, then A is positive semidefinite (negative semidefinite).

MATRIX INVERSE

Let A and B be two n×n matrices. The matrix A^{-1}, a unique inverse of A, exists if and only if A is nonsingular. A^{-1} has the property that $A^{-1}A = AA^{-1} = I$. If A^{-1} and B^{-1} exist, $(AB)^{-1} = B^{-1}A^{-1}$.

POSITIVE DEFINITE TEST AND MATRIX INVERSION

The inversion method of FORTRAN subroutine INVERT (listed on page 375) is a straightforward Gauss-Jordan method [P6]. The diagonal entries are taken in succession to be the pivotal entries; if a given pivotal entry is nonpositive, the A matrix is not positive definite (this test is directly related to Sylvester's determinant test, each pivotal entry being the ratio of appropriate determinants). The inversion requires N^2 divisions, $N^3 - N^2$ multiplications, and $N^3 - N^2$ subtractions.

D MATRIX OPERATIONS

```
      SUBROUTINE INVERT(AA,A,AI,N,N2,IPOS,EP)
C  IN THE CALLING PROGRAM,AA MUST BE DIMENSIONED
C  N×2N, AND BOTH A AND AI MUST BE DIMENSIONED
C  N×N. 'A' IS THE MATRIX TO BE INVERTED.
C  N2 = 2*N IS SET IN THE CALLING ROUTINE.
C  EP, AN ERROR PARAMETER, MUST BE ASSIGNED
C  (TYPICALLY, EP = 1.E-10).
C  IF INVERT GIVES IPOS = 1,'A' IS POSITIVE DEFINITE.
C  IF INVERT GIVES IPOS =-1,'A' IS NOT POSITIVE
C  DEFINITE.
C  IF IPOS = 1 OR -1, THE INVERSE OF 'A' IS PLACED
C  IN AI.
C  A PIVOTAL ENTRY WITH VALUE
C  BETWEEN EP AND -EP CAUSES IPOS = 0 TO BE ASSIGNED,
C  IN WHICH CASE THE INVERSION PROCESS IS NOT
C  COMPLETED.
      DIMENSION AA(N,N2),A(N,N),AI(N,N)
      IPOS=1
      DO 10 I=1,N
      DO 10 J=1,N
      AA(I,J)=A(I,J)
   10 AA(I,J+N) = 0.
      DO 20 I=1,N
   20 AA(I,I+N) = 1.
      DO 80 I=1,N
      IF (AA(I,I).GT.EP) GO TO 40
      IF (AA(I,I).LT.-EP) GO TO 30
      IPOS=0
      RETURN
   30 IPOS= -1
   40 DO 50 K=I+1,N+I
   50 AA(I,K) = AA(I,K)/AA(I,I)
      DO 60 J=1,N
      IF (J.EQ.I) GO TO 60
      DO 55 K=I+1,N+I
   55 AA(J,K) = AA(J,K) - AA(I,K)*AA(J,I)
   60 CONTINUE
   80 CONTINUE
      DO 100 I=1,N
      DO 100 J=1,N
  100 AI(I,J) = AA(I,J+N)
      RETURN
      END
```

LINEAR EQUATION SOLUTION

Given is an m×n matrix A, an n×1 vector x of unknowns, and an m×1 vector b of numbers. The equation

$$Ax = b \tag{D-9}$$

is solvable for x if and only if the rank of A equals the rank of the partitioned matrix [A, b].

In the case where A is n×n, and if A^{-1} exists, the solution of (D-9) is $x = A^{-1}b$.

LEAST SQUARES APPROXIMATION

If (D-9) is replaced by

$$Ax = b + \varepsilon \tag{D-10}$$

where ε is an m×1 vector, a pseudo solution x^* for x can always be obtained on the basis that $\varepsilon'\varepsilon$ is minimized with respect to x. That is

$$(Ax^* - b)'(Ax^* - b) = \min_{x}(Ax - b)'(Ax - b) \tag{D-11}$$

from which, x^* can be shown to satisfy

$$A'Ax^* = A'b \tag{D-12}$$

D MATRIX OPERATIONS

INNER PRODUCT AND OUTER PRODUCT

Given are n×1 vectors x and y. The product x´y is a scalar and is the inner product of x and y. The product xy´ is an n×n matrix and is an outer product of x and y.

NONTRIVIAL VECTOR

Any vector having one or more nonzero element is a nontrivial vector.

ORTHOGONAL VECTORS

The n×1 nontrivial vectors x and y are said to be mutually orthogonal if x´y = 0.

EUCLIDEAN NORM

The Euclidean norm of an n×1 vector x of real entries is denoted by $||x||$ and is defined by

$$||x|| = (x´x)^{\frac{1}{2}} \tag{D-13}$$

ORTHOGONAL PROJECTION IN E^n

Let U be the set defined as follows:

$$U \triangleq \{z | z = \alpha_1 r_1 + \alpha_2 r_2 + \cdots + \alpha_m r_m\} \quad (D-14)$$

where $r_1, r_2, \ldots, r_m$ are given n×1 vectors, with n > m, and the α_i's are scalars that can assume any real values. Corresponding to any vector x contained in n-dimensional Euclidean space E^n, there exists a vector $x_p \in U$ with the property that

$$(x - x_p)'z = 0 \quad (D-15)$$

for all $z \in U$; that is, $(x - x_p)$ is orthogonal to all vectors contained in U. It can be shown that (D-15) is equivalent to

$$\|x - x_p\| = \min_{z \in U} \|x - z\| \quad (D-16)$$

PROJECTION OF x ONTO y

Let x and y be n×1 vectors in E^n. The projection of x onto y satisfies (D-16); here,

$$\|x - x_p\| = \min_{\alpha} \|x - \alpha y\| \quad (D-17)$$

which is satisfied by $x_p = [(x'y)/(y'y)]y$.

GRADIENT VECTOR

Let $f(x)$ be a scalar function of n variables $x = \{x_1, x_2, \ldots, x_n\}$. If $f(x)$ is of class C^1,

D MATRIX OPERATIONS

the $n \times 1$ vector $\partial f(x)/\partial x = [\partial f/x_1 \quad \partial f/\partial x_2 \quad \cdots \quad \partial f/\partial x_n]' = \nabla f$ is the gradient of f with respect to x.

VECTOR DERIVATIVE OF A ROW MATRIX

Let $p = [p_1(x) \quad p_2(x) \quad \cdots \quad p_m(x)]$ where each $p_i(x)$ is a scalar function of class C^1 with respect to $x = \{x_1, x_2, \ldots, x_n\}$. The derivative of p with respect to x is the $n \times m$ matrix:

$$\frac{\partial p}{\partial x} = \begin{bmatrix} \frac{\partial p_1}{\partial x} & \frac{\partial p_2}{\partial x} & \cdots & \frac{\partial p_m}{\partial x} \end{bmatrix} \qquad (D-18)$$

VECTOR DERIVATIVE OF A COLUMN MATRIX

Let $q = [q_1(x) \quad q_2(x) \quad \cdots \quad q_m(x)]'$ where each $q_i(x)$ is a scalar function of class C^1 with respect to $x = \{x_1, x_2, \ldots, x_n\}$. The derivative of q with respect to x is the $m \times n$ matrix

$$\frac{\partial q}{\partial x} = \begin{bmatrix} \frac{\partial q_1}{\partial x} & \frac{\partial q_2}{\partial x} & \cdots & \frac{\partial q_m}{\partial x} \end{bmatrix}', \quad m \neq 1 \qquad (D-19)$$

It follows that $(\partial q/\partial x)' = \partial(q')/\partial x$. The matrix of (D-19) is called the <u>Jacobian</u> of q with respect to x.

VECTOR DERIVATIVE OF AN INNER PRODUCT

Let p and q be m×1 vector functions of class C^1 with respect to the vector x. Then

$$\partial(p'q)/\partial x = (\partial p/\partial x)'q + (\partial q/\partial x)'p \qquad (D-20)$$

COMMON MATRIX DERIVATIVES

Let A be an n×n matrix independent of x. Then $\partial(x'A)/\partial x = A$ and $\partial(x'Ax)/\partial x = Ax + A'x$.

EIGENVALUES or CHARACTERISTIC ROOTS

Let A be an n×n matrix of reals. The eigenvalues or characteristic roots of A are the roots of the nth-order polynomial equation

$$|\lambda I - A| = (\lambda - \lambda_1)(\lambda - \lambda_2) \cdots (\lambda - \lambda_n) = 0 \qquad (D-21)$$

where λ is a complex variable, and I is an identity matrix. If A is also symmetric and positive definite, all λ_i roots can be shown to be real and positive.

EIGENVECTOR OR CHARACTERISTIC VECTOR

Only eigenvalues λ_i's of A satisfy

D MATRIX OPERATIONS

$$Az_i = \lambda_i z_i \qquad (D-22)$$

for some nontrivial vectors z_i's. The z_i's are eigenvectors or characteristic vectors of A.

SIMILAR MATRICES

Given are n×n matrices A and B. If

$$B = P^{-1}AP \qquad (D-23)$$

for some nonsingular matrix P, A and B are similar matrices. Similar matrices have the same eigenvalues.

MATRICES SIMILAR TO DIAGONAL MATRICES

Given is an n×n matrix A and an n×n diagonal matrix D. The following properties are useful.

1. If A is similar to D, the diagonal entries of D are the eigenvalues of A.

2. If the eigenvalues of A are distinct, $A = PDP^{-1}$ where the jth column of P is an eigenvector that corresponds to the eigenvalue in the jth column of D.

3. If A is a symmetric matrix of reals, n linearly independent and mutually orthogonal eigenvectors exist.

4. If the orthogonal eigenvectors of property number 3 above are normalized to unity magnitude, and are used as the columns of P, then $P^{-1} = P'$.

5. If conditions numbered 3 and 4 above apply, the nonsingular linear transformation $x = Py$ can be used to simplify the quadratic form $x'Ax$ to the noninteracting form

$$y'Dy = \sum_{i=1}^{n} \lambda_i y_i^2 \qquad (D-24)$$

APPENDIX E

UNCONSTRAINED SEARCH

E.1 INTRODUCTION

For comparison purposes, sequential search techniques can be divided into two categories: 1) techniques that utilize derivatives of the problem functions, and 2) those that do not require derivatives of the problem functions. In general, the best sequential methods which utilize first derivatives (gradient information) are more efficient than those that do not; the search techniques employed by the multiplier algorithm of this text utilize first derivatives in generating the search directions.

The techniques presented in this appendix are invariably found in the literature as they apply to the minimization problem. In keeping with the theme of this text, these techniques are presented here in forms appropriate to the maximization problem (max f = -min(-f)).

E.2 UNCONSTRAINED SEARCH

Sequential unconstrained maximization procedures locate a local maximum of the objective function $f(x)$ as the limit of the sequence $\{x^k\}$, $k = 0, 1, 2, \ldots$, where x^0 yields an initial estimate of the maximum, and for each $k \geq 0$, x^{k+1} yields an approximation to the maximum of f with respect to variations along the straight line through x^k in some specified direction r^k.

A desirable property for an iterative search method to possess is that of <u>quadratic convergence</u>; meaning that for quadratic functions, the maximum will be located exactly, apart from round-off errors, within a finite number of iterations. In general, an unconstrained performance measure, resulting from the transformation of the constrained NLP into an unconstrained problem, is not quadratic -- but for an analytic function $f(x)$, the following argument applies. In a Taylor series expansion of $f(x)$, expanded about a local maximum

E.2 UNCONSTRAINED SEARCH

point x^*, the second-order (quadratic) terms dominate higher-order terms as search points x^k's approach x^*. The function then appears more quadratic in nature, and the rate of convergence generally increases.

In regions remote from a maximum, iterative search methods often account for the curvature of the function and are thereby able to deal with complex situations which are created, for example, when active constraints are incorporated with penalty functions. Information about the local behavior of the function is most efficiently used when iterative search methods calculate each new direction of search as part of the iteration cycle.

In the remainder of this appendix, we investigate the properties of several gradient methods used for generating search directions. Common to most of these gradient techniques is the recursion (1-14) given here in the form

$$x^{k+1} = x^k + \rho H^k g^k \qquad (E-1)$$

where x^{k+1} is the new value of x obtained from a variation ρ along the direction $H^k g^k$. H^k is an n×n matrix, g^k is the gradient of f evaluated at

x^k, and ρ is a real number. The methods based on (E-1) differ in the way that ρ and H^k are selected.

E.3 STEEPEST ASCENT

In the method of steepest ascent, H^k in (E-1) is set to the identity matrix I for all k. At the maximum point of each unidirectional search, the next direction is set equal to the gradient at that point, which is the direction of greatest incremental increase in f(x).

Evaluating f(x) at a general point defined by (E-1) with H = I, we have

$$f(x^{k+1}) = f(x^k + \rho g(x^k)) \triangleq y(\rho) \qquad (E-2)$$

which forms a parametric function in ρ. The first derivative of $y(\rho)$ with respect to ρ is

$$\frac{\partial y}{\partial \rho} = g(x^{k+1})'g(x^k) \qquad (E-3)$$

If $\|g(x^k)\| \neq 0$, $\partial y/\partial \rho$ is positive at $\rho = 0$ because $g(x^k)'g(x^k) > 0$. Therefore, an increase in f is generated at every step if $\rho > 0$ is sufficiently small. If ρ is too small, an increase is guaranteed, but convergence may be very slow, and if ρ is too large, f generally decreases. The greatest

E.3 UNCONSTRAINED SEARCH

increase in f occurs from (E-3) at a value of ρ where $\partial y/\partial \rho = 0$. This optimum value is found or a close approximation to it is found by operating on (E-2) with a unidirectional search such as the one outlined in Section 5.4.

Convergence, although most certain with the above method, is usually slow and relatively inefficient, as oscillatory search directions are often generated. Curry [C10] and others note that the convergence rate of this method is affected by scaling of the x_i variables. For example, in E^n, if the function being maximized is spherical with the maximum at the center of the sphere, the maximum can be found in one iteration, because the gradient is normal to the spherical surface and points toward the center. If a scale change for any coordinate is made, however, the surface becomes ellipsoidal, the gradient generally does not point toward the center, and maximization requires more than one iteration, resulting in a poorer rate of convergence.

E.4 NEWTON SEARCH

Suppose that in the vicinity of the current search point x^k, $f(x^{k+1})$ can be adequately represented by the truncated Taylor series

$$f(x^{k+1}) = f(x^k) + \Delta x' g(x^k) + \tfrac{1}{2}\Delta x' A^k \Delta x \qquad (E-4)$$

where $\Delta x = x^{k+1} - x^k$, and A^k is the Hessian of f evaluated at x^k. The Δx that maximize $f(x^{k+1})$ of (E-4) must satisfy

$$A^k \Delta x = -g(x^k) \qquad (E-5)$$

If $(A^k)^{-1}$ exists, (E-5) can be rearranged as

$$x^{k+1} = x^k + (-A^k)^{-1} g^k \qquad (E-6)$$

which is in the form of (E-1) with $\rho = 1$ and $H^k = (-A^k)^{-1}$. If the function to be maximized is quadratic and has a well-defined maximum, A^k is constant, and equation (E-6) gives the required maximum point x^* in one iteration from any point x^k.

The basic Newton Search is given by (E-6). It has the property of quadratic convergence, but requires the evaluation of first and second derivatives and the inversion of a matrix for every iteration. When x^k is near a local maximum point of a nonquadratic function, such that f can be

E.4 UNCONSTRAINED SEARCH

approximated by (E-4), and if $-A^k$ is positive definite, rapid convergence is assured. But if x^k is not in the vicinity of a local maximum, such that the higher-order terms not shown in (E-4) dominate the right-hand side of the Taylor series approximation, $-A^k$ may not be positive definite, and the basic Newton Search may actually diverge. Also, if A^k is a singular matrix, such that the determinant $|-A^k|$ of $-A^k$ is zero, the inverse of $-A^k$ does not exist, and (E-6) cannot be applied. Even when $|-A^k| \neq 0$, it may be so close to zero that $-A^k$ is ill-conditioned, leading to erroneous results.

E.5 CONJUGATE GRADIENT METHODS

Conjugate direction methods can be regarded as being somewhat intermediate between the method of steepest ascent and Newton's method [L7]. They are motivated by the desire to accelerate the slow convergence associated with steepest ascent while avoiding the information requirements associated with the evaluation, storage, and inversion of the Hessian as required by Newton's method.

The original conjugate gradient method was developed by Hestenes and Stiefel [H5], to solve a set of simultaneous linear equations having a symmetric positive-definite matrix of coefficients. The equivalence of the linear problem

$$Ax = b \qquad (E-7)$$

and the maximization of a quadratic function can be seen from equation (E-5) which is linear in Δx with solution

$$\Delta x = (-A^k)^{-1} g(x^k) \qquad (E-8)$$

Let us define and state some properties of conjugate direction vectors.

DEFINITION: [Conjugate vectors] Let H be a positive-definite matrix. A vector v^1 is said to be conjugate (with respect to H) to a vector v^2 if v^1 and v^2 satisfy

$$(v^1)'Hv^2 = 0 \qquad (E-9)$$

When v^1 and v^2 are interpreted as direction vectors in E^n and satisfy (E-9), they are said to be H-conjugate directions. For example, in the method of steepest ascent, the vectors $g(x^k)$ and $g(x^{k+1})$ are conjugate with respect to the identity matrix I, because the optimum value of ρ found for each

E.5 UNCONSTRAINED SEARCH

search gives $\partial y/\partial \rho = 0$ in equation (E-3) from which

$$g(x^{k+1})'Ig(x^k) = g(x^{k+1})'g(x^k) = 0 \qquad \text{(E-10)}$$

Consider the following properties of conjugate directions, which are given in [P4] as theorems with proofs. If H is an n×n positive-definite matrix, then:

a) If n nontrivial directions r^i, $i = 1, 2, \ldots, n$, are known to be H conjugate to one another, then these r^i's are linearly independent.

b) When $f(x)$ is a quadratic with A equal to the Hessian, and when nontrivial directions r^i, $i = 0, 1, \ldots, n-1$ are mutually conjugate with respect to A, the exact maximum point x^* of $f(x)$ can be obtained by a sequence of n unidirectional searches starting at x^0. The x^k are determined from the iteration formula

$$f(x^{k+1}) = \max_{\rho} f(x^k + \rho r^k), \qquad \text{(E-11)}$$
$$k = 0, 1, 2, \ldots, n-1$$

with the final result being $x^n = x^*$.

Property b demonstrates that the method of sequential unidirectional searches is quadratically convergent when using a set of A-conjugate search directions on a quadratic function with Hessian A.

Equation (E-9) gives a definite criterion that conjugate search directions must satisfy. In some problem formulations of the NLP, only the problem functions and their gradients are defined numerically; the Hessian is not explicitly available. Additionally, the f(x) to be maximized is not quadratic in general, and if x is far from a local maximum, other relations must be used to estimate conjugate directions. Conjugate directions based on linear independence can be generated without using the gradient of f(x). Methods of this sort are discussed in [P4, P11, Z1]. Several well-known methods for generating conjugate directions when the Hessian is not available are derived in [L7] and in the references cited there. One such method that generates conjugate search directions from gradient information at consecutive search points is the method of Fletcher and Reeves.

E.6   METHOD OF FLETCHER AND REEVES

In the conjugate gradient method of Fletcher and Reeves [F11], the conjugate directions for the

E.6 UNCONSTRAINED SEARCH

general nonquadratic problem are generated as follows:

$$r^i = g^i \quad \text{for} \quad i = j(n + 1),$$
$$j = 0, 1, 2, \ldots \quad \text{(E-12)}$$

and

$$r^{i+1} = g^{i+1} + \frac{(g^{i+1})'(g^{i+1})}{(g^i)'(g^i)} r^i,$$

otherwise (E-13)

where g^{i+1} equals the gradient of $f(x)$ at the point x^{i+1}, and where x^{i+1} is determined from

$$f(x^{i+1}) = \max_{\rho} f(x^i + \rho r^i) \quad \text{(E-14)}$$

Convergence of this method is established by noting that the algorithm is restarted by a search in the steepest ascent direction every n + 1 iterations. (Actually, restarts somewhat less frequent generally give better results.) Because searches in the conjugate directions do not decrease the objective, and usually increase it, convergence is assured [L7]. Thus the cyclical restart of the algorithm is important for convergence, because in the general nonlinear case, one can not guarantee that the directions generated by (E-13) alone are ascent directions. (The direction-generation algorithm should be restarted when a nonascent direction is generated.) The method has the property of quadratic convergence, and it has a storage advantage: no n×n H matrix need be generated and stored. Computationally, however, it has not proved to be as effective as some quasi-Newton methods.

E.7 QUASI-NEWTON METHODS

The quasi-Newton methods, also known as variable-metric methods, are considered to be the most sophisticated methods for solving the general unconstrained problem. The basic idea of the quasi-Newton methods is to iteratively use an improved approximation to the inverse Hessian instead of the true inverse as in Newton's method. The improved approximations are generated from information obtained during the ascent process. These methods thus gain some of the advantages of Newton's method while using only first-order information about the function. These methods are based upon the recursion given in (E-1). The improved

E.7 UNCONSTRAINED SEARCH

approximations H^i to the inverse Hessian $(-A)^{-1}$ required in (E-8) are inferred from gradient information at previous interations and are updated as additional gradient information become available. The update is performed such that

$$H^{i+1} \Delta g^i = \Delta x^i \qquad (E-15)$$

where $\Delta x^i = x^{i+1} - x^i$, $\Delta g^i = g^{i+1} - g^i$, g^{i+1} is the gradient of $f(x)$ at the point x^{i+1}, and x^{i+1} is found from the recursion

$$f(x^{i+1}) = \max_{\rho} f(x^i + \rho H^i g^i) \qquad (E-16)$$

The updating criterion (E-15), often referred to as the quasi-Newton condition, is motivated by the following: if the objective is quadratic and if $H^{i+1} = (-A)^{-1}$ for some i, then (E-15) yields (E-8).

E.8 THE DFP METHOD

The first quasi-Newton algorithm, and to date still one of the most powerful and sophisticated of the gradient methods, is Fletcher and Powell's [F10] development of Davidon's [D2] Variable Metric Method, the Davidon-Fletcher-Powell (DFP) method.

In this method H^i is updated according to

$$H^{i+1} = H^i - \frac{(\Delta x^i)(\Delta x^i)'}{(\Delta x^i)'(\Delta g^i)}$$

$$- \frac{H^i(\Delta g^i)(\Delta g^i)'H^i}{(\Delta g^i)'H^i(\Delta g^i)} \qquad (E-17)$$

where Δx^i and Δg^i are as in (E-15) and x^{i+1} is found from (E-16).

The method possesses three properties that form the basis for its highly successful convergence characteristics:

a) Given that H^0 is positive definite, each H^i generated by (E-17) will be positive definite.

b) If the objective is quadratic and $H^0 = I$, the search directions are identical to those of the conjugate gradient method, and convergence to the solution is obtained in n steps.

c) If the objective is quadratic, and if convergence to the solution requires n steps, the nth approximation H^n is equal to the inverse Hessian.

The field of quasi-Newton methods has been a very active research area. Many of the published contributions are surveyed in [A2, B11, H6, M5, P13, P14]. Several researchers propose alternate

E.8 UNCONSTRAINED SEARCH

updating formulae possessing some or all of the three properties of the DFP algorithm. Some of them [A2, B7, G5] introduce unifying approaches that lead to general classes of such formulae. A very general class of such updating formulae is considered by Huang [H7] who derives a general family of algorithms (requiring a unidirectional search) that have the second property given above [O3]. Oren [O2] surveys many of these general classes and their motivations, and gives yet another approach based on geometric considerations. His approach adopts the view that the most important feature in the quasi-Newton algorithms is generating good approximations of the inverse Hessian, rather than the generation of conjugate directions. His motivation stems from the fact that variable metric methods tend to be sensitive to factors that weaken their approximation to the conjugate gradient method. For example [L7, O3], the performance of the DFP method tends to be less efficient as the accuracy of the line search decreases. It is also noted to be somewhat sensitive to scaling of the variables in the objective function and to uniform scaling of the variables through multiplication by a scaler. Poor scaling may cause the update matrix H^i to become singular due to roundoff error, causing the algorithm to fail unless corrective action is taken to reinitialize the algorithm. Proper initial scaling may improve the performance of the DFP algorithm, but there is no provision to do automatic scaling while the algorithm is executing.

In the original form of the DFP method (and in many other quasi-Newton methods that update the inverse Hessian approximation), it is implicit that no predetermined resets of H be made; the algorithm consists of starting with an initial approximation for H and successively improving it throughout the iterative process [L7]. Global convergence, however, has not been proven for this procedure and computational experiments and investigations of the convergence properties of this method suggest that the DFP method operated in this way can fail (see Section 7.4). It is suggested, therefore, that practical criteria be incorporated to reinitialize the algorithm; such criteria are presented in Chapter 5.

E.9 SELF-SCALING VARIABLE-METRIC ALGORITHM

In a series of works [O1, O3, O4], Luenberger and Oren consider a new criterion for comparing the convergence properties of variable metric algorithms. They focus on stepwise descent (ascent) properties to derive bounds on the rate of decrease (increase) in the function value at each iterative step. By applying (E-16) to a quadratic function f with the maximum at x^* and with H^i drawn from a uniformly positive-definite family, they find that the following condition holds:

$$f(x^{i+1}) - f(x^*) \le \left[\frac{K(T^i) - 1}{K(T^i) + 1}\right]^2 (f(x^i) - f(x^*)) \tag{E-18}$$

where $K(T^i)$ denotes the <u>condition number</u> of the matrix $T^i = (-A)^{\frac{1}{2}} H^i (-A)^{\frac{1}{2}}$; that is, $K(T^i)$ is the ratio of the largest to the smallest eigenvalue of T^i.

Consider convergence in the light of (E-18). For steepest ascent, $H^i = I$ for all i, so $T^i = (-A)$, and convergence is linear with a rate bounded in terms of $K(-A)$. For Newton's method, $H^i = (-A)^{-1}$ so $T^i = I$ and $f(x^{i+1}) = f(x^*)$ for any i, meaning that the maximum is reached in one step. It follows that, to guarantee good convergence at each step, the term $\left[\dfrac{K(T^i) - 1}{K(T^i) + 1}\right]^2$ referred to as the single-step convergence factor should be made as small as possible, and therefore, one should strive to make $K(T^i)$ approach 1 (that is, to minimize $K(T^i)$). Furthermore, if $K(T^i) > K(-A)$ for some i, the convergence rates on these steps may be worse than for steepest ascent. This was shown [L6, O3] to be true for the DFP method in a quadratic case when a fixed error was introduced in the unidirectional search, causing poor eigenvalue structure in the algorithm and destroying conjugacy.

Luenberger and Oren use (E-18) as a basis for algorithm development. They introduce variable coefficients to rescale the objective function at each iteration and thereby form a new class of variable-metric algorithms. Effective scaling can be implemented by restricting the parameters in a two parameter family of variable-metric algorithms. Conditions are derived [O3] for these parameters that guarantee monotonic improvement in the

E.9 UNCONSTRAINED SEARCH

single-step convergence rate. The conditions were obtained by analyzing the eigenvalue structure of the associated inverse Hessian approximations. Their development leads to a two parameter family of algorithms given in the following equivalent form:

$$H^{i+1} = \left[H^i - \frac{H^i(\Delta g^i)(\Delta g^i)'H^i}{(\Delta g^i)'H^i(\Delta g^i)} + \theta^i v^i (v^i)' \right] \gamma^i$$

$$- \frac{(\Delta x^i)(\Delta x^i)'}{(\Delta x^i)'(\Delta g^i)} \qquad \text{(E-19)}$$

where

$$v^i = [(\Delta g^i)'H^i(\Delta g^i)]^{\frac{1}{2}} \left[\frac{H^i(\Delta g^i)}{(\Delta g^i)'H^i(\Delta g^i)} \right.$$

$$\left. - \frac{(\Delta x^i)}{(\Delta x^i)'(\Delta g^i)} \right] \qquad \text{(E-20)}$$

and the parameters satisfy

$$\gamma^i > 0, \quad \theta^i \geq 0 \qquad \text{(E-21)}$$

The vectors (Δx^i), (Δg^i) are as in (E-15), and x^{i+1} is found from (E-16).

This class is equivalent to a subset of Huang's family [H7]. In the above form, it is guaranteed that (E-15) holds for any choice of γ^i and θ^i. This property implies that at least in one direction the inverse Hessian appoximation agrees with the inverse Hessian so that their spectrums of eigenvalues overlap. Thus, $\rho = 1$ can always be used as an initial estimate for ρ in the unidirectional search. Also the optimum steps will tend to unity, thereby eliminating the need for a unidirectional search. The positivity restrictions on θ^i and γ^i are necessary to guarantee positive-definiteness of the matrices H^i.

The conceptual framework for self-scaling algorithms is to systematically implement the idea of improving the performance of variable-metric methods by scaling the objective function. Updating formulae of the type given in (E-19), (E-20), and (E-21) are said to be self scaling if for any fixed positive (negative) definite quadratic with Hessian A, the parameters γ^i and θ^i are automatically selected such that $K(T^{i+1}) \leq K(T^i)$ for every i independent of the updating vector Δx^i. ($T^i = (-A)^{\frac{1}{2}} H^i (-A)^{\frac{1}{2}}$, $\Delta g^i = A\Delta x^i$). Particular cases of this algorithm that use self-scaling updating formulae are referred to as Self-Scaling Variable-Metric (SSVM) algorithms. The key result for SSVM

E.9 UNCONSTRAINED SEARCH

algorithms is as follows: if $\theta \in [0, 1]$ and if γ^i is chosen such that $1/\gamma^i$ is in the interval spanned by the eigenvalues of T^i (i.e., $\lambda_1^i \leq 1/\gamma^i \leq \lambda_n^i$), then $K(T^{i+1}) \leq K(T^i)$. The practicality of such an algorithm therefore depends on the availability of a procedure to derive suitable γ^i from readily available information without evaluating the eigenvalues of each T^i.

One method of selecting the scaling factor γ^i is given in [04]. An equivalent form is

$$\gamma^i = \frac{(\Delta x^i)'(\Delta g^i)}{(\Delta g^i)' H^i (\Delta g^i)} (\phi - 1)$$

$$- \frac{(g^i)'(\Delta x^i)}{(g^i)' H^i (\Delta g^i)} \phi \qquad (E-22)$$

where $\phi \in [0, 1]$.

The equation set (E-19), (E-20), (E-21), and (E-22) along with (E-16) form a general family of SSVM's that have the following properties:

a) The inverse Hessian approximations are positive definite.

b) For an n-dimensional quadratic function, the directions of search are conjugate and the algorithms converge in n steps.

c) The single-step convergence rate decreases monotonically when the algorithm is applied to a quadratic function.

d) The algorithms are invariant under scaling of the objective function or uniform scaling of the variables.

For the quadratic case, if γ^0 is generated by (E-22) and if $\gamma^i = 1$ for all $i > 0$, it is noted in [O4] that the resulting algorithm maintains the property that $H^n = (-A)^{-1}$. We call this property Property 1 (P1). On the other hand, adjusting γ^i for every step as in (E-22), P1 is lost in the quadratic case. But by allowing γ^i to vary, a monotonic improvement in the single-step convergence rate is ensured, which is implied by a monotonic decrease in $K(T^i)$. We call this property Property 2 (P2). The implications of trading P1 ($H^n = (-A)^{-1}$) for P2 ($K(T^{i+1}) \leq K(T^i)$) should be understood.

According to Oren [O4] the trading of P1 for P2 is meaningless in the quadratic case where $H^0 = I$ and where exact line search is performed because both methods are conjugate gradient methods and reach the maximum in n steps, regardless of the

E.9 UNCONSTRAINED SEARCH 405

fact that the $(n + 1)^{th}$ step is a Newton step for P1. In order to understand the meaning of trading these two properties, we must consider practical cases and the consequences of perturbations that destroy the conjugacy and n-step convergence. Such cases come about from higher-order terms in the objective, roundoff errors, and inaccuracies in the unidirectional search. It may be argued in such cases that P1 may still tend to provide a good step every $(n + 1)^{th}$ iteration, even though the progress at intermediate steps may be poor. For example, if the algorithm jams for n steps in the neighborhood of a given point, H^{n+1} may yet be a good approximation of the inverse Hessian, yielding an approximate Newton step. P2 on the other hand will show progress at each step at least as good as steepest ascent.

In view of the above argument, an advantage may be anticipated for P1 in the case of a difficult function with few variables. Here a good $(n + 1)^{th}$ step will compensate for poor intermediate steps. As the number of variables becomes larger, the importance of P1 decreases while that of P2 increases. This can be explained by the possibility of having many ineffective steps with Pl (at least when searching in areas remote from a local maximum) and by the cumulative effect of good intermediate steps provided by P2. This becomes increasingly important when the number n of variables is large, in which case the number of iterations required to obtain a reasonable approximation to the solution is expected to be a low multiple of n.

Oren [O5] performed a parametric study on ϕ and θ to determine the effect of these parameters on the performance of the SSVM algorithm. He also investigated a switching strategy for varying these parameters, and studied the effect of reduced frequency of unidirectional search by using a controlled parameter. He found that the algorithm performed best for the choice of $\theta = \frac{1}{2}$ and $\phi = 1$, independent of unidirectional search precision. No theoretical explanation was given to justify this empirical conclusion. For the unconstrained problem, he found the SSVM algorithm to be competitive with the DFP algorithm for low dimensioned

E.9 UNCONSTRAINED SEARCH

functions and superior for larger problems. The results for the proposed switching strategy were not as good as those for the above fixed parameter selection. No results were indicated as to what one might expect for SSVM performance on the general constrained NLP.

E.10 COMMENT

The multiplier algorithm LPNLP of this text incorporates both the DFP method and the SSVM method as options for generating search directions through the update of H^i. Either method also has the option of being restarted ($H^0 = I$) after a multiplier update phase (provided that at least n unidirectional searches have been performed since the last restart). Thus, a total of 4 possible modes for search direction generation are available. To reduce possible confusion for the user, SSVM is implemented in all cases with $\theta = \frac{1}{2}$ and $\phi = 1$, as appears reasonable from Oren's results [O5].

An economical unidirectional search (see Section 5.4) is employed for all modes of the algorithm. It uses a 'window' scheme that in some cases (especially in the end search) allows a single fixed step of $\rho = 1$ to satisfy the ascent requirements of the search. The search is conducted over $\rho \geq 0$ only; thus, ascent directions are required. If any of the 4 modes fail to generate an ascent direction, the direction defaults to the gradient direction, yielding a restart with the first direction being one of the steepest ascent.

The results of Chapter 6 confirm Oren's results for unconstrained problems, with SSVM performance being generally better than that of DFP. For constrained problems, however, the methods reverse roles, with DFP outperforming SSVM. Intuitive explanations of this result are as follows:

> a) In the development of the SSVM algorithm, no general formula is known for generating γ^i that guarantees γ to be optimal. When the eigenvalues of T^i exhibit small separation, the performance of the SSVM algorithm is relatively insensitive to variations of $\gamma \in [1/\lambda^n, 1/\lambda^1]$. When applied to the augmented Lagrangian of the general constrained problem, however, cases arise in which there are large separations

E.10 UNCONSTRAINED SEARCH

of the eigenvalues. Thus, there is a wide allowable range for the selection of γ.

b) Some of the advantages of the self-scaling property P2 may be lost for the constrained problem. At the beginning of the LPNLP maximization process, the status of the constraints may change quite rapidly from free to locked-on constraints and vice versa. In this situation, substantial changes can occur in the characteristics of the augmented Lagrangian and in the eigenvalue structure of the Hessian.

c) The updating procedure, changing weights and multipliers in the augmented Lagrangian, tends to devaluate accumulated scaling information generated by SSVM.

REFERENCES

A1 Abadie, J., "On the Kuhn-Tucker Theorem," Chapter 2 of Nonlinear Programming, J. Abadie (Editor), North-Holland Publishing Company, Amsterdam, 1967.

A2 Adachi, N., "On Variable-Metric Algorithms," Journal of Optimization Theory and Applications, Vol. 7, pp. 391-410, June 1971.

A3 Alsac, O., and B. Stott, "Optimal Load Flow with Steady-State Security," IEEE Transactions on Power Apparatus and Systems, Vol. PAS-93, pp. 745-751, May/June 1974.

A4 Armacost, R.L., and A.V. Fiacco, "Computational Experience in Sensitivity Analysis for Nonlinear Programming," Mathematical Programming, Vol. 6, pp. 301-326, June 1974.

A5 Arrow, K.J., F.J. Gould, and S.M. Howe, "A General Saddle Point Result for Constrained Optimization," Institute of Statistics Mimeo Series No. 774, University of North Carolina, Chapel Hill, September 1971. (see also, Mathematical Programming, Vol. 5, pp. 225-234, Oct. 1973.)

REFERENCES

A6 Arrow, K.J., and L. Hurwicz, "Reduction of Constrained Maxima to Saddle Point Problems," in *Third Berkeley Symposium on Mathematical Statistics and Probability*, J. Neyman (Editor), University of California Press, Berkeley, 1956.

A7 _____ and R.M. Solow, "Gradient Methods for Constrained Maxima, with Weakened Assumptions," in *Studies in Linear and Nonlinear Programming*, K. Arrow, L. Hurwicz, and H. Uzawa (Editors), Stanford University Press, Stanford, 1958.

A8 Asaadi, J., "A Computational Comparison of Some Nonlinear Programs," *Mathematical Programming*, Vol. 4, pp. 144-154, April 1973.

A9 Avriel, M., M.J. Rijckaert, and D.J. Wilde (Editors), *Optimization and Design* (from a summer school on the impact of optimization theory on technological design), Prentice-Hall, Englewood Cliffs, N.J., 1973.

B1 Bandler, J.W., and R.E. Seviora, "Current Trends in Network Optimization," *IEEE Transactions on Microwave Theory and Techniques*, Vol. MIT-18, pp. 1159-1170, Dec. 1970.

B2 Beale, E.M.L., "Numerical Methods," Chapter 7 of *Nonlinear Programming*, J. Abadie (Editor), North-Holland Publishing Company, Amsterdam, 1967.

B3 _____, *Mathematical Programming in Practice*, Sir Isaac Pitman and Sons Ltd., London, 1968.

B4 Bellman, R., *Introduction to Matrix Analysis*, McGraw-Hill, New York, 1960.

B5 Beltrami, E.J., "A Constructive Proof of the Kuhn-Tucker Multiplier Rule," *Journal of Mathematical Analysis and Applications*, Vol. 26, pp. 297-306, May 1969.

B6 Biggs, M.C., "Minimization Algorithms Making Use of Non-Quadratic Properties of the Objective Function," *Journal of the Institute of Mathematics and Its Applications*, Vol. 8, pp. 315-327, Dec. 1971.

B7 Box, M.J., "A Comparison of Several Current Optimization Methods, and the Use of Transformations in Constrained Problems," Computer Journal, Vol. 9, pp. 67-77, May 1966.

B8 _____, D. Davies, and W.H. Swann, Nonlinear Optimization Techniques, (IMPERIAL Chemical Industries LIMITED MONOGRAPH No. 5), Oliver and Royd, Edinburgh 1, 1969.

B9 Bracken, J., and G.P. McCormick, Selected Applications of Nonlinear Programming, Wiley, New York, 1968.

B10 Brameller, A., and K.L. Lo, "A Review of Minimization Techniques with Reference to Power System Engineering," in Real-Time Control of Electrical Power Systems, Symposium Proceedings, Edmund Handschin (Editor), Brown, Boveri and Company, Ltd., Baden, Switzerland, 1971, Elsevier Publishing Company, Amsterdam, 1972.

B11 Broyden, C.G., "Quasi-Newton Methods and Their Application to Function Minimization," Mathematics of Computation, Vol. 21, pp. 368-381, July 1967.

C1 Cagnon, C.R., R.H. Hicks, S.L.S. Jacoby, and J.S. Kowalik, "A Nonlinear Programming Approach to a Very Large Hydroelectric System Optimization," Mathematical Programming, Vol. 6, pp. 28-41, Feb. 1974.

C2 Carroll, C.W., "The Created-Response-Surface Technique for Optimizing Nonlinear Restrained Systems," Operations Research, Vol. 9, pp. 169-184, March/April 1961.

C3 Charnes, A, and W.W. Cooper, "Chance-Constrained Programming," Management Science, Vol. 6, pp. 73-79, Oct. 1959.

C4 _____ and _____, "Chance Constraints and Normal Deviates," Journal of the American Statistical Association, Vol. 57, pp. 134-148, March 1962.

REFERENCES

C5 Colville, A.R., "A Comparative Study on Nonlinear Programming Codes," IBM New York Scientific Center Report No. 320-2949, June 1968.

C6 Courant, R., "Variational Methods for the Solution of Problems of Equilibrium and Vibrations," Bulletin of the American Math Society, Vol. 49, pp. 1-23, Jan. 1943.

C7 _____, Differential and Integral Calculus, Vols. I and II, Wiley-Interscience, New York, 1936, Translation by E.J. McShane.

C8 Coxeter, H.S.M., "The Golden Section, Phyllotaxis, and Wythoff's Game," Scripta Mathematica, Vol. 19, pp. 135-143, June-Sept. 1953.

C9 Crabill, T.B., J.P. Evans, and F.J. Gould, "An Example of an Ill-Conditioned NLP Problem," Mathematical Programming, Vol. 1, pp. 113-116, Oct. 1971.

C10 Curry, H.B., "The Method of Steepest Descent for Nonlinear Minimization Problems," Quarterly of Applied Mathematics, Vol. 2, pp. 258-261, Oct. 1944.

D1 Dantzig, G.B., "Maximization of a Linear Function of Variables Subject to Linear Inequalities," Chapter 21 of Activity Analysis of Production and Allocation, T.C. Koopmans (Editor), Cowles Commission Monograph No. 13, Wiley, New York, 1951.

D2 Davidon, W.C., "Variable Metric Method for Minimization," Atomic Energy Commission Research and Development Report, ANL-5990, 1959.

D3 Davies, D, and W.H. Swann, "Review of Constrained Optimization," Chapter 12 of Optimization, R. Fletcher (Editor), Academic Press, London, 1969.

D4 Davis, R.H., and P.D. Roberts, "Method of Conjugate Gradients Applied to Self-Adaptive Digital Control Systems," Proceedings of the Institution of Electrical Engineers (London), Vol. 115, pp. 562-571, April 1968.

D5 Debreu, G., "Definite and Semidefinite Quadratic Forms," Econometrica, Vol. 20, pp. 295-300, April 1952.

D6 Dennis, J.B., Mathematical Programming and Electrical Networks, Massachusetts Institute of Technology Press, 1959.

D7 Dinkel, J.J., and G.A. Kochenberger, "On a Cofferdam Design Optimization," Mathematical Programming, Vol. 6, pp. 114-116, Feb. 1974.

D8 Dommel, H., and W. Tinney, "Optimal Power Flow Solutions," IEEE Transactions on Power Apparatus and Systems, Vol. PAS-87, pp. 1866-1876, Oct. 1968.

D9 Duffin, R.J., E.L. Peterson, and C. Zener, Geometric Programming - Theory and Application, Wiley, New York, 1967.

E1 El-Abiad, A.H., and F.J. Jaimes, "A Method for Optimum Scheduling of Power and Voltage Magnitude," IEEE Transactions on Power Apparatus and Systems, Vol. PAS-88, pp. 413-422, April 1969.

E2 Evans, J.P., F.J. Gould, and J.W. Tolle, "Exact Penalty Functions in Nonlinear Programming," Mathematical Programming, Vol. 4, pp. 72-97, Feb. 1973.

F1 Faddeeva, V.N., Computational Methods of Linear Algebra, Translated from the Russian version by C.D. Benster, Dover Publications, New York, 1959.

F2 Farkas, J., "Uber die Theorie der Einfachen Ungleichungen, J. Reine Angew. Math., Vol. 124, pp. 1-27, 1901.

F3 Fiacco, A.V., and G.P. McCormick, "The Sequential Unconstrained Minimization Technique for Nonlinear Programming, A primal-Dual Method," Management Science, Vol. 10, pp. 360-366, Jan. 1964.

REFERENCES

F4 Fiacco, A.V., and G.P. McCormick, "Computational Algorithm for the Sequential Unconstrained Minimization Technique for Nonlinear Programming," Management Science, Vol. 10, pp. 601-617, July 1964.

F5 _____ and _____, "Extensions of SUMT for Nonlinear Programming; Equality Constraints and Extrapolation," Management Science, Vol. 12, pp. 816-828, July 1966.

F6 _____ and _____, Nonlinear Programming: Sequential Unconstrained Minimization Techniques, Wiley, New York, 1968.

F7 Fletcher, R. (Editor), Optimization, Academic Press, London, 1969.

F8 _____, "A Class of Methods for Nonlinear Programming with Termination and Convergence Properties," Chapter 6 of Integer and Nonlinear Programming, J. Abadie (Editor), North-Holland Publishing Company, Amsterdam, 1970.

F9 _____ and S.A. Lill, "A Class of Methods for Nonlinear Programming, II; Computational Experience," in Nonlinear Programming, J.B. Rosen, O.L. Mangasarian and K. Ritter (Editors), Academic Press, pp. 67-92, 1971.

F10 _____ and M.J.D. Powell, "A Rapidly Convergent Descent Method for Minimization," The Computer Journal, Vol. 7, pp. 163-168, July 1963.

F11 _____ and C.M. Reeves, "Function Minimization by Conjugate Gradients," The Computer Journal, Vol. 7, pp. 149-154, July 1964.

F12 Forsythe, G.E., "Computing Constrained Minima with Lagrange Multipliers," Journal of the Society for Industrial and Applied Mathematics, Vol. 3, pp. 173-178, Dec. 1955.

F13 Fox, K.A., J.K. Sengupta, and G.V.L. Narasimham (Editors), Economic Models, Estimation and Risk Programming, Springer-Verlag, New York, 1969.

G1   Gallagher, R.H., and O.C. Zienkiewicz (Editors), *Optimum Structural Design*, Wiley, New York, 1973.

G2   Geoffrion, A.M. (Editor), *Perspectives on Optimization*, Addison-Wesley, Reading, Mass., 1972.

G3   Gottfried, B.S., and J. Weisman, *Introduction to Optimization Theory*, Prentice-Hall, Inc., Englewood Cliffs, New Jersey, 1973.

G4   Gould, F.J., and J.W. Tolle, "A Necessary and Sufficient Qualification for Constrained Optimization," *SIAM Journal on Applied Mathematics*, Vol. 20, pp. 164-172, March 1971.

G5   Greenstadt, J., "Variations on Variable Metric Methods," *Mathematics of Computation*, Vol. 24, pp. 1-22, Jan. 1970.

H1   Haarhoff, P.C., and J.D. Buys, "A New Method for the Optimization of a Nonlinear Function to Nonlinear Constraints," *The Computer Journal*, Vol. 13, pp. 178-184, May 1970.

H2   Hadley, G., *Nonlinear and Dynamic Programming*, Addison-Wesley, Reading, Mass., 1964.

H3   Hestenes, M.R., "Multiplier and Gradient Methods," *Journal of Optimization Theory and Applications*, Vol. 4, pp. 303-320, Nov. 1969.

H4   _____, "Multiplier and Gradient Methods," in *Computing Methods in Optimization Problems - 2*, pp. 143-163, L.A. Zadeh, L.W. Neustadt, and A.V. Balakrishman (Editors), Academic Press, New York, 1969.

H5   _____ and E. Stiefel, "Methods of Conjugate Gradients for Solving Linear Systems," *Journal of Research of the National Bureau of Standards*, Vol. 49, pp. 409-436, Dec. 1952.

H6   Himmelblau, D.M., "A Uniform Evaluation of Unconstrained Techniques," pp. 69-97 in *Numerical Methods for Nonlinear Optimization*, F.A. Lootsma (Editor), Academic Press, London, 1972.

REFERENCES

H7 Huang, H.Y., "Unified Approach to Quadratically Convergent Algorithms for Function Minimization," <u>Journal of Optimization Theory and Applications</u>, Vol. 5, pp. 405-423, June 1970.

J1 John, F., "Extreme Problems with Inequalities as Side Conditions," in <u>Studies and Essays</u>, Courant Anniversary Volume, K.O. Friedrichs, O.E. Neugebauer, and J.J. Stoker (Editors), Wiley, New York, pp. 187-204, 1948.

K1 Kavlie, D., and J. Moe, "Application of Nonlinear Programming to Optimum Grillage Design with Non-Convex Sets of Variables," <u>International Journal for Numerical Methods in Engineering</u>, Vol. 1, pp. 351-378, Oct.-Dec. 1969.

K2 Keefer, D.L., and B.S. Gottfried, "Differential Constraint Scaling in Penalty Function Optimization," Abstract, <u>Bulletin Operations Research Society of America</u>, Vol. 18, Sup. 2, p. B-184, Oct. 1970.

K3 King, R.P., "Necessary and Sufficient Conditions for Inequality Constrained Extreme Values," <u>Industrial and Engineering Chemistry Fundamentals</u>, Vol. 5, pp. 484-489, Nov. 1966.

K4 Kort, B.W., and D.P. Bertsekas, "Multiplier Methods for Convex Programming," <u>Proceedings of the IEEE Decision and Control Conference</u>, San Diego, California, Dec. 1973.

K5 _____ and _____, "Combined Primal-Dual and Penalty Methods for Convex Programming," Bell Telephone Laboratories Report, Holmdel, New Jersey, Sept. 1974.

K6 Kuhn, H.W., and A.W. Tucker, "Nonlinear Programming," in <u>Proceedings of the Second Berkeley Symposium on Mathematical Statistics and Probability</u>, J. Neyman (Editor), University of California Press, Berkeley, pp. 481-493, 1951.

K7　Kwakernaak, H., and R.C.W. Strijbos, "Extremization of Functions with Equality Constraints," *Mathematical Programming*, Vol. 2, pp. 279-295, June 1972.

L1　Lootsma, F.A., "Hessian Matrices of Penalty Functions for Solving Constrained Optimization Problems," *Philips Research Reports*, Vol. 24, pp. 322-330, Aug. 1969.

L2　_____, "Penalty-Function Performances of Several Unconstrained-Minimization Techniques," *Philips Research Reports*, Vol. 27, pp. 358-385, Aug. 1972.

L3　Lowe, M.J., "Programming Instructions for Subroutine ALGOR," Electronics Research Laboratory Report No. 2773, Montana State University, Bozeman, Montana, Dec. 1973.

L4　_____, "Nonlinear Programming: Augmented Lagrangian Techniques for Constrained Minimization," Ph.D. Thesis, Montana State University, Bozeman, Montana, March 1974.

L5　Luenberger, D.G., *Optimization by Vector Space Methods*, Wiley, New York, 1969.

L6　_____, "Convergence Rate of a Penalty-Function Scheme," *Journal of Optimization Theory and Applications*, Vol. 7, pp. 39-51, Jan. 1971.

L7　_____, *Introduction to Linear and Nonlinear Programming*, Addison-Wesley, Reading, Mass., 1973.

M1　McCormick, G.P., "Penalty Function Versus Non-Penalty Function Methods for Constrained Nonlinear Programming Problems," *Mathematical Programming*, Vol. 1, pp. 217-238, Nov. 1971.

M2　Mangasarian, O.L., "Unconstrained Lagrangians in Nonlinear Programming," Computer Sciences Technical Report No. 174, University of Wisconsin, Madison, March 1973.

REFERENCES

M3 Miele, A., E.E. Cragg, R.R. Iyer, and A.V. Levy, "Use of the Augmented Penalty Function in Mathematical Programming Problems, Part 1," *Journal of Optimization Theory and Applications*, Vol. 8, pp. 115-130, Aug. 1971.

M4 _____, _____ and A.V. Levy, "Use of the Augmented Penalty Function in Mathematical Programming Problems, Part 2," *Journal of Optimization Theory and Applications*, Vol. 8, pp. 131-153, Aug. 1971.

M5 _____, H.Y. Huang, and J.W. Contrell, "Gradient Methods in Mathematical Programming, Part 1, Review of Previous Techniques" *Aero - Astronautics Report No. 55*, Rice University, Houston, Texas, 1969.

M6 _____, P.E. Moseley, and E.E. Cragg, "Numerical Experiments on Hestenes' Method of Multipliers for Mathematical Programming Problems," *Aero - Astronautics Report No. 85*, Rice University, Houston, Texas, 1971.

M7 _____, _____ and _____, "A Modification of the Method of Multipliers for Mathematical Programming Problems," *Aero - Astronautics Report No. 86*, Rice University, Houston, Texas, 1971.

M8 _____, _____, A.V. Levy, and G.M. Coggins, "On the Method of Multipliers for Mathematical Programming Problems," *Journal of Optimization Theory and Application*, Vol. 10, pp. 1-33, July 1972.

M9 Misra, K.B., and M.D. Ljubojević, "Optimal Reliability Design of a System: A New Look," *IEEE Transactions on Reliability*, Vol. R-22, pp. 255-258, Dec. 1973.

M10 Murray, W., "Ill-Conditioning in Barrier and Penalty Functions Arising in Constrained Nonlinear Programming," *Proc. of the Sixth International Symposium on Mathematical Programming*, Princeton University, Aug. 13-20, 1967.

N1   Neghabat, F., and R.M. Stark, "A Cofferdam Design Optimization," *Mathematical Programming*, Vol. 3, pp. 263-275, Dec. 1972.

O1   Oren, S.S., "Self-Scaling Variable Metric Algorithms Without Line Search for Unconstrained Minimization," *Mathematics of Computation*, Vol. 27, pp. 873-885, Oct. 1973.

O2   _____, "Quasi-Newton Algorithms: Approaches and Motivations," *Proceedings of IEEE Decision and Control Conference*, San Diego, California, Dec. 1973.

O3   _____ and D.G. Luenberger, "Self-Scaling Variable Metric (SSVM) Algorithms, Part I: Criteria and Sufficient Conditions for Scaling a Class of Algorithms," *Management Science*, Vol. 20, pp. 845-862, Jan. 1974.

O4   _____, "Self-Scaling Variable Metric (SSVM) Algorithms, Part II: Implementation and Experiments," *Management Science*, Vol. 20, pp. 863-874, Jan. 1974.

O5   _____, "On the Selection of Parameters in Self-Scaling Variable Metric Algorithms," *Mathematical Programming*, Vol. 7, pp. 351-367, Dec. 1974.

P1   Pearson, J.D., "Variable Metric Methods of Minimization," *The Computer Journal*, Vol. 12, pp. 171-178, May 1969.

P2   Perlis, S., *Theory of Matrices*, Addison-Wesley, Reading, Mass., 1958 (3rd printing).

P3   Pierre, D.A., "Sensitivity Measures for Optimally Selected Parameters," *Proceedings of the IEEE*, Vol. 54, pp. 321-322, Feb. 1966.

P4   _____, *Optimization Theory with Applications*, Wiley, New York, 1969.

REFERENCES

P5 Pierre, D.A., "Nonlinear Programming: Research Goals and Preliminary Results," *Research Memorandum S10*, Electronics Research Laboratory, Montana State University, Bozeman, Montana, Dec. 1971.

P6 _____, "A Nongradient Minimization Algorithm Having Parallel Structure, with Implications for an Array Computer," *Computers and Electrical Engineering*, Pergamon Press, Vol. 1, pp. 3-21, June 1973.

P7 _____, "Multiplier Algorithms for Nonlinear Programming," Presented at the 8th International Symposium on Mathematical Programming, Aug. 27-31, 1973.

P8 _____, "Two Multiplier Algorithms for Nonlinear Programming," September, 1973. (Revised version of a paper presented at the 8th International Symposium on Mathematical Programming, Aug. 27-31, 1973.)

P9 Pinal, J.F., and K.A. Roberts, "Tolerance Assignment in Linear Networks Using Nonlinear Programming," *IEEE Transactions on Circuit Theory*, Vol. CT-19, pp. 475-479, Sept. 1972.

P10 Porcelli, G., and K.A. Fegley, "Optimal Design of Digitally Compensated Systems by Quadratic Programming," *Journal of the Franklin Institute*, Vol. 282, pp. 303-317, Nov. 1966.

P11 Powell, M.J.D., "An Efficient Method for Finding the Minimum of a Function of Several Variables without Calculating Derivatives," *The Computer Journal*, Vol. 7, pp. 155-162, July 1964.

P12 _____, "A Method for Nonlinear Constraints in Minimization Problems," in *Optimization*, R. Fletcher (Editor), Academic Press, New York, 1969.

P13 _____, "A Survey of Numerical Methods for Unconstrained Optimization," *SIAM Review*, Vol. 12, pp. 79-97, Jan. 1970.

P14 Powell, M.J.D., "Recent Advances in Unconstrained Optimization," Mathematical Programming, Vol. 1, pp. 26-57, Oct. 1971.

R1 Rockafellar, R.T., "New Applications of Duality in Convex Programming," presented at the 7th International Symposium on Mathematical Programming (the Hague, 1970), published in the Proceedings of the 4th Conference on Probability (Brasov, Romania, 1971).

R2 _____, "Penalty Methods and Augmented Lagrangians in Nonlinear Programming," Proc. 5th IFIP Conference on Optimization Techniques, Rome, May 1973.

R3 _____, "A Counter Example," personal correspondence to D.A. Pierre, Aug. 31, 1973.

R4 _____, "A Dual Approach to Solving Nonlinear Programming Problems by Unconstrained Optimization," Mathematical Programming, Vol. 5, pp. 354-373, Dec. 1973.

R5 _____, "Augmented Lagrange Multiplier Functions and Duality in Nonconvex Programming," SIAM Journal on Control, Vol. 12, pp. 268-285, May 1974.

R6 _____, "The Multiplier Method of Hestenes and Powell Applied to Convex Programming," Journal of Optimization Theory and Applications, to be published.

R7 Rosen, J.B., "The Gradient Projection Method for Nonlinear Programming, Part 1: Linear Constraints," Journal of the Society for Industrial and Applied Mathematics, Vol. 9, pp. 514-532, Dec. 1961.

R8 _____ and S. Suzuki, "Construction of Nonlinear Programming Test Problems," Communications of the Association for Computing Machinery, Vol. 8, pg. 113, Feb. 1965.

REFERENCES

R9   Rosenbrock, H.H., "An Automatic Method for Finding the Greatest or the Least Value of a Function," *The Computer Journal*, Vol. 3, pp. 175-184, Oct. 1960.

S1   Sasson, A.M., "Nonlinear Programming Solutions for Load Flow, Minimum Loss, and Economic Dispatching Problems," *IEEE Transactions on Power Apparatus and Systems*, Vol. PAS-88, pp. 399-409, April 1969.

S2   Sayama, H., Y. Kameyama, H. Nakayama, and Y. Sawaragi, "The Generalized Lagrangian Functions for Mathematical Programming Problems," Report 55, Institute for Systems Design and Optimization, Kansas State University, Manhattan, Kansas, Feb. 1974.

S3   Shanno, D.F., "Conditioning of Quasi-Newton Methods for Function Minimization," *Mathematics of Computation*, Vol. 24, pp. 647-656, July 1970.

S4   Siddall, J.N., *Analytical Decision Making in Engineering Design*, Prentice-Hall, Englewood Cliffs, New Jersey, 1972.

S5   Spunt, L., *Optimum Structural Design*, Prentice-Hall, Englewood Cliffs, New Jersey, 1971.

S6   Steenbrink, P.A., *Optimization of Transport Networks*, Wiley, New York, 1974.

T1   Tabak, D., and B.C. Kuo, *Optimal Control by Mathematical Programming*, Prentice-Hall, Englewood Cliffs, New Jersey, 1971.

T2   Taha, H.A., *Operations Research, An Introduction*, The Macmillan Company, New York, 1971.

T3   Taylor, F.J., and R.J. Molepske, "Optimal Filter Design via Mathematical Programming," *IEEE Transactions on Systems, Man, and Cybernetics*, Vol. SMC-3, pp. 382-389, July 1973.

T4   Thorbjornsen, A.R., and S.W. Director, "Computer Aided Tolerance Assignment for Linear Circuits with Correlated Elements," <u>IEEE Transactions on Circuit Theory</u>, Vol. CT-20, pp. 518-523, Sept. 1973.

V1   Vajda, S., "Nonlinear Programming and Duality," Chapter 1 of <u>Nonlinear Programming</u>, J. Abadie (Editor), North-Holland Publishing Company, Amsterdam, 1967.

V2   van de Panne, C., and W. Popp, "Minimum-Cost Cattle Feed under Probalistic Protein Constraints," <u>Management Science</u>, Vol. 9, pp. 405-430, April 1963.

V3   Varaiya, P.P., <u>Notes on Optimization</u>, Van Nostrand Reinhold Co., New York, 1972.

W1   Witte, B.F.W., and W.R. Holst, "Two New Direct Minimum Search Procedures for Functions of Several Variables," <u>AFIPS Conference Proceedings</u>, American Federation of Information Processing Societies, Spring Joint Computer Conference, Vol. 25, pp. 195-209, 1964.

W2   Wolfe, P., "Methods of Nonlinear Programming," Chapter 6 of <u>Nonlinear Programming</u>, J. Abadie (Editor), North-Holland Publishing Company, Amsterdam, 1967.

W3   _____, "Explicit Solution of an Optimization Problem," <u>Mathematical Programming</u>, Vol. 2, pp. 258-260, April 1972.

W4   Wong, K.P., "Decentralized Planning by Vertical Decomposition of an Economic System: A Nonlinear Approach," Ph.D. Thesis, University of Birmingham, Birmingham, England, Sept. 1970.

Z1   Zangwill, W.I., "Minimizing a Function without Calculating Derivatives," <u>The Computer Journal</u>, Vol. 10, pp. 293-296, Nov. 1967.

Z2   _____, <u>Nonlinear Programming: A Unified Approach</u>, Prentice-Hall, Englewood Cliffs, New Jersey, 1969.

REFERENCES

Z3   Zoutendijk, G., *Methods of Feasible Directions*, Elsevier Publishing Company, Amsterdam, 1960.

Z4   Zwart, P., "Nonlinear Programming: Global Use of the Lagrangian," *Journal of Optimization Theory and Applications*, Vol. 6, pp. 150-160, Aug. 1970.

AUTHOR INDEX

Abadie, J., 35, 102 [A1]
Adachi, N., 396-397 [A2]
Alsac, O., 261 [A3]
Armacost, R.L., 257 [A4]
Arrow, K.J., 38, 40, 365 [A5-A7]
Asaadi, J., 228, 237-239, 241, 244-245 [A8]
Avriel, M., 260 [A9]
Balakrishman, A.V., 416 [H4]
Bandler, J.W., 262 [B1]
Beale, E.M.L., 117, 235 [B2, B3]
Bellman, R., 368 [B4]
Beltrami, E.J., 36, 173 [B5]
Benster, C.D., 414 [F1]
Bertsekas, D.P., 40, 366 [K4, K5]
Biggs, M.C., 221-222, 254 [B6]
Box, M.J., 35, 254, 397 [B7, B8]
Bracken, J., 260, 267 [B9]
Brameller, A., 261 [B10]
Broyden, C.G., 396 [B11]

Buys, J.D., 39 [H1]
Cagnon, C.R., 261 [C1]
Carroll, C.W., 31, 32 [C2]
Charnes, A., 267, 273 [C3, C4]
Coggins, G.M., 39 [M8]
Colville, A.R., 220, 226, 257 [C5]
Contrell, J.W., 396 [M5]
Cooper, W.W., 267, 273 [C3, C4]
Courant, R., 31, 37, 109 [C6, C7]
Coxeter, H.S.M., 43 [C8]
Crabill, T.B., 255 [C9]
Cragg, E.E., 39 [M3, M4, M6, M7]
Curry, H.B., 387 [C10]
Dantzig, G.B., 2 [D1]
Davidon, W.C., 192, 202, 217, 395 [D2]
Davies, D., 29, 35 [B8, D3]
Davis, R.H., 261 [D4]
Debreu, G., 140 [D5]

AUTHOR INDEX

Dennis, J.B., 262 [D6]
Dinkel, J.J., 262 [D7]
Director, S.W., 262 [T4]
Dommel, H., 261 [D8]
Duffin, R.J., 260 [D9]
El-Abiad, A.H., 261 [E1]
Evans, J.P., 40, 255 [C9, E2]
Faddeeva, V.N., 368 [F1]
Farkas, J., 83 [F2]
Feder, D., 255
Fegley, K.A., 261 [P10]
Fiacco, A.V., 30, 33, 35, 36, 83, 102, 234, 256, 257 [A4, F3-F6]
Fletcher, R., 39, 40, 192, 392, 395 [F7-F11]
Forsythe, G.E., 171 [F12]
Fox, K.A., 260 [F13]
Friedrichs, K.O., 417 [J1]
Gallagher, R.H., 261 [G1]
Geoffrion, A.M., 157 [G2]
Gottfried, B.S., 35, 46 [K2]
Gould, F.J., 40, 102, 255, 365 [C9, E2, G4]
Greenstadt, J., 397 [G5]
Haarhoff, P.C., 39 [H1]
Hadley, G., 260 [H2]
Handschin, E., 412 [B10]
Hestenes, M.R., 38-39, 256, 365, 390 [H3-H5]
Hicks, R.H., 261 [C1]
Himmelblau, D.M., 396 [H6]
Holst, W.R., 253-254 [W1]
Howe, S.M., 40, 365 [A5]
Huang, H.Y., 396, 397, 401 [H7, M5]
Hurwicz, L., 38, 365 [A6]
Iyer, R.R., 39 [M3]
Jacoby, S.L.S., 261 [C1]
Jaimes, F.J., 261 [E1]
John, F., 37, 78 [J1]
Kameyama, Y., 40, 366 [S2]
Kavlie, D., 262 [K1]
Keefer, D.L., 35 [K2]
King, R.P., 38 [K3]

Kochenberger, G.A., 262 [D7]
Koopmans, T.C., 413 [D1]
Kort, B.W., 40, 366 [K4, K5]
Kowalik, J.S., 261 [C1]
Kuhn, H.W., 2, 37, 74, 78 [K6]
Kuo, B.C., 261 [T1]
Kwakernaak, H., 40 [K7]
Lagrange, J.L., 363
Levy, A.V., 39 [M3, M4, M8]
Lill, S.A., 39 [F9]
Ljubojević, M.D., 293 [M9]
Lo, K.L., 261 [B10]
Lootsma, F.A., 35, 36, 255 [L1, L2]
Lowe, M.J., 41, 127, 267, 367 [L3, L4]
Luenberger, D.G., 35, 36, 166, 168, 257, 368, 389, 392-393, 397-400 [L5-L7, O3]
McCormick, G.P., 30, 33, 35, 36, 83, 102, 234, 256, 260, 267 [B9, F3-F6, M1]
McShane, E.J., 413 [C7]
Mangasarian, O.L., 40, 365-366 [M2]
Miele, A., 39, 396 [M3-M8]
Misra, K.B., 293 [M9]
Moe, J., 262 [K1]
Molepske, R.J., 262 [T3]
Moseley, P.E., 39 [M6-M8]
Murray, W., 36 [M10]
Nakayama, H., 40, 366 [S2]
Narasimham, G.V.L., 260 [F13]
Neghabat, F., 262 [N1]

Neugebauer, O.E., 417 [J1]
Neustadt, L.W., 416 [H4]
Neyman, J., 411, 417 [A6, K6]
Oren, S.S., 219, 221-223, 226-227, 397, 399-400, 403-404, 406-408 [O1-O5]
Pearson, J.D., 36 [P1]
Perlis, S., 368 [P2]
Peterson, E.L., 260 [D9]
Pierre, D.A., 35, 41, 78, 113, 127, 131, 202, 228, 232-233, 281, 367, 374, 391-392 [P3-P8]
Pinel, J.F., 262 [P9]
Popp, W., 267 [V2]
Porcelli, G., 261 [P10]
Powell, M.J.D., 38-39, 192, 220, 236, 392, 395, 396 [F10, P11-P14]
Reeves, C.M., 392 [F11]
Rijckaert, M.J., 260 [A9]
Ritter, K., 415 [F9]
Roberts, K.A., 262 [P9]
Roberts, P.D., 261 [D4]
Rockafellar, R.T., 40, 136, 246 [R1-R6]
Rosen, J.B., 2, 234 [R7, R8]
Rosenbrock, H.H., 71, 219 [R9]
Sasson, A.M., 261 [S1]
Sawaragi, Y., 40, 366 [S2]
Sayama, H., 40, 366 [S2]
Sengupta, J.K., 260 [F13]
Seviora, R.E., 262 [B1]
Shanno, D.F., 254 [S3]
Siddall, J.N., 260 [S4]
Solow, R.M., 38, 365 [A7]
Spunt, L., 261 [S5]
Stark, R.M., 262 [N1]

Steenbrink, P.A. 260 [S6]
Stiefel, E., 256, 390 [H5]
Stoker, J.J., 417 [J1]
Stott, B., 261 [A3]
Strijbos, R.C.W., 40 [K7]
Suzuki, S., 234 [R8]
Swann, W.H., 29, 35 [B8, D3]
Tabak, D., 261 [T1]
Taha, H.A., 260, 291 [T2]
Taylor, F.J., 262 [T3]
Thorbjornsen, A.R., 262 [T4]
Tinney, W., 261 [D8]
Tolle, J.W., 40, 102 [E2, G4]
Tucker, A.W., 2, 37, 74, 78 [K6]
Uzawa, H., 411 [A7]
Vajda, S., 157 [V1]
van de Panne, C., 267 [V2]
Varaiya, P.P., 117 [V3]
Weisman, J., 46 [G3]
Wilde, D.J., 260 [A9]
Witte, B.F.W., 253-254 [W1]
Wolfe, P., 30, 35, 255 [W2, W3]
Wong, K.P., 237-239, 241 [W4]
Zadeh, L.A., 416 [H4]
Zangwill, W.I., 35, 259, 392 [Z1, Z2]
Zener, C., 260 [D9]
Zienkiewicz, O.C., 261 [G1]
Zoutendijk, G., 83 [Z3]
Zwart, P., 254 [Z4]

SUBJECT INDEX

Abnormal case, 74-75
Algebraic equation
  solution, 48
Applications,
  typical, 259-262
Architect's problem,
  43
Area-volume problem,
  64-67
  an augmented La-
    grangian for, 134-
    136
  duality analysis of,
    162-165
  sensitivity analysis,
    114-117, 123
Around-the-world
  problem, 228-231
Attribute costs, 117
Augmented function, 31
Augmented Lagrangian,
  40
  gradient of, 130-131
  L_a of this text,
    128-129
  methods, 31
  reduced form of, 156

Augmented Lagrangian,
  strict local maximum
    of, 134, 139
  theorems concerning
    L_a, 131-132, 137-
    138, 140
Banana function, 219
Beale's problem, 235
Biggs' function,
  $N = 3$, 221
  $N = 5$, 222
Bounds,
  FORTRAN implementa-
    tion, 314-315
  upper and lower,
    6, 216
C^0, C^1, C^2, 53
C_a, 94, 129
C_b, 129
Calling program, 295
  example of, 320
Chance-constrained
  problem, 267-279
  an investment prob-
    lem, 44
Characteristic roots,
  380

Circle inscribed in a
  triangle, 46
  sensitivity aspect,
    126
Circuits problem,
  279-288
Class C^k, 54
Concave functions, 18
  on sums of, 20-21
Concave functions,
  particular Lagrangian,
    107
Cone,
  convex polyhedral, 85
Conjugate gradient
  methods, 389-394
Constraint basis,
  148-149
Constraint breakthrough,
  151-152
Constraint qualifica-
  tion, 102
Constraints,
  active, 68
  an inequality con-
    straint case,
    67-71
  inactive, 68
  nonzero gradient
    assumptions, 52
  one equality con-
    straint case,
    59-67
Container problem,
  262-267
Continuity, 52-53
Contours,
  equimagnitude, 63
Control applications,
  260-261
Control cards,
  316-317
Convergence,
  finite convergence
    for linear case,
    152-153

Convergence factor,
  single-step, 400-401
Convergence rate,
  399-400
  and scale change,
    387
  as a function of
    weights, 168-169
Convergence test,
  in FORTRAN, 306
Convex functions, 18
  examples of, 47
  on sums of, 20-21
Convex polyhedron, 101
Convex programming, 37
  inequality con-
    strained, 40
Convex set, 17
  closed, 18
  defined by concave
    functions, 23
  defined by convex
    functions, 22-23
  defined by linear
    equalities, 26
  examples of, 47
  intersection of, 24
Conveyor problem, 45,
  292
  with friction, 292
Cost function, 4
Cube function, 253
Cubic approximation,
  217
Cycle, 195, 214
Data formats, 302-316
Data set,
  example, 320
Design-center values,
  112, 115
Determinant, 372
DFP method, 192, 216,
  395-398
  compared to SSVM,
    250, 408-409

SUBJECT INDEX

DFP method,
  re accuracy of line
    search, 397, 400
  re resets, 398
  re scaling, 397-398
  results of, 223-226,
    242-243, 250
Diagonal matrix, 370
Differentiability,
  classes of, 53
Dimensions used in
  LPNLP, 298-299
Double function, 254
Double precision,
  on the use of,
    317-318
Dual function,
  local, 157
Dual methods, 31, 39
Dual problem, 160,
  example, 162-165
Duality,
  local dual assump-
    tions, 154-156
Duality theorem, 161
Dynamic programming, 41
Economic problems, 260
Eigenvalues, 380
Eigenvector, 380-381
Electrical networks,
  262
Equality constraints, 5
Euclidean norm, 16, 51,
  377

Euclidean space E^n, 4
Farkas lemma, 83-86
Feasible points, 5, 32
Feasible Region, 90,
  extended, 53-54
  strictly feasible, 32
Feed ration problem,
  267-279
Filter design, 262,
  a digital filter,
    293-294

Finite convergence,
  for linear problem,
    152-154
Flags for LPNLP,
  303-306
Fletcher-Reeves
  search, 392-394
Free constraint, 149
Functions,
  examples of one
    variable, 7-14
  (see also particular
    function types)
FXNS,
  example of, 319
Gauss-Jordan method,
  374
Geometrical pro-
  gramming, 260
Golden section ratio,
  43
GRAD,
  subroutine example,
    319
Gradient, 55,
  case where nonzero
    assumption fails,
    106-107
  nonzero assumption,
    52
Gradient component,
  constant versus
    nonconstant,
    300-302
Gradient search
  methods, 385-386
Gradient vector,
  6, 378-379
Heart function, 254
Hessian matrix, 55,
  ill-conditioning of,
    36
  of a dual function,
    158
Hyperplane, 60
Identity matrix, 370

Inequality constraints,
  5-7
Inner product,
  derivative of, 380
Integer Programming,
  41
Interest rate problem,
  44
Investment problem,
  44, 291
Jacobian matrix, 379
Kuhn-Tucker conditions,
  37,
  first-order
    conditions, 87-88
  second-order
    conditions, 94-96,
    102
L (see Lagrangian)
L_a (see Augmented
  Lagrangian)
Lagrange multipliers,
  38,
  for one equality
    constraint case,
    61, 64
Lagrange Multipliers,
  generalized, 38
  relation to penalty
    terms, 36
  versus substitution,
    105
Lagrangian,
  a concave case, 107
  a deficiency of, 127
  for an inequality
    constraint case,
    70-73
  for one equality
    constraint, 61
  for the general case,
    78-79
  modified forms of,
    363-367
  penalized, 39

Lagrangian,
  (see also Augmented
    Lagrangian)
Lagrangian function,
  37,
  modified, 38
Least Squares,
  a problem, 44, 290
  for linear
    equations, 376
Line in E^n, 199, 216
Line search
  (see unidirectional
    search)
Linear combination,
  371
Linear equation
  solution, 376, 390
Linear dependence,
  89
Linear independence,
  89, 371
Linear problem,
  atypical cases, 101
  re finite
    convergence, 152
Linear programming,
  1, 41
Linear test problem,
  232
Linear transformation,
  382
Locked-on status,
  149-150
LPNLP,
  instructions for
    using, 295-318
  results, 225-227,
    242-253
  typical output,
    321-325
MacLaurin series, 55
Management problem,
  sensitivity analysis,
    118, 122

SUBJECT INDEX

Marginal productivities, 105
Matrix, 368-382,
  addition, 369
  condition number of, 399
  derivatives, 380
  ill-conditioned, 389
  inverse, 374-375
  multiplication, 369
  partitioned, 371
  positive definite, 59
  rank of, 371
  similar ones, 381
  symmetric, 370
  transpose, 51, 370
Maximum,
  absolute, 7
  constrained local, 51-52
  global for convex programming case, 6, 26-27
  local, 16
  multiple solution case, 255
  of an augmented Lagrangian, 131-132, 137-138, 140
  strict constrained local, 52
  with one equality constraint, 59-67
  (see also, Optimum)
Minimum,
  local, 17
  (see also, Optimum)
Minimization,
  equivalence to maximization, 4
Monotonic function, 14
Multiplier Algorithm,
  an overview of, 144-152
  comparison basis, 218

Multiplier Algorithm,
  duality concepts, 165-166
  essential structure for, 175-177
  finite convergence for general linear case, 152-153
  maximization phase, 146-147
  program notation, 187-189
  subroutine connections for, 178
  update phase, 147
  user routines, 179-180
Multiplier functions, 364-367
Multiplier methods, 31, 38,
  for convex programming, 40
Multiplier update rule, and local duality, 165-168
Multipliers,
  for general case, 78, 87
  geometrical interpretations, 63, 72
  nonunique case of, 254
  update rules for, 147-148
Necessary conditions,
  for general case, 81, 83
  for inequality constraint case, 69-70
  second-order, 96
Negative definite, 59, 373
Neighborhood, 53

Newton method,
  166, 388-389
NLP, 1
  an ill-conditioned
    one, 255-256
  FORTRAN form,
    295-302
  functions, 5
  in a standard form,
    4-7, 50-51
  problem statement,
    5-7
  unconstrained, 3
Noninteracting form,
  382
Normal case, 74, 76-77
Normal distribution,
  268, 274
Normal solution, 103
Normal vector, 60
Objective function, 4
Optimal point, 5
Optimum,
  first-order condition
    for unconstrained
    case, 57
  global for linear
    problems, 2
  second-order
    conditions for
    unconstrained case,
    58
  unconstrained local,
    56
  (see also, Maximum)
OPTKOV, 244-245
Orthogonal projection,
  377-378
Parametric form,
  55, 92, 386
Parcel problem, 71-73,
  98-100
  sensitivity of,
    123-124
Penalty function,
  exterior-point
    function, 35
Penalty function,
  interior-point
    function, 34
  mixed case, 34
Penalty function
  methods, 31-36
Penalty weights,
  (see Weights)
Performance measure,
  4
Positive definite,
  59, 373
POWCON, 244-245
Powell's function, 220
Power output problem,
  107
Power systems, 261
Power transfer
  problem, 45
Present worth, 44
Primal problem, 160
Production problem,
  43, 105-106, 290,
  292, 294
Projection,
  of a vector, 85
Quadratic approxima-
  tion, 202-204
Quadratic convergence,
  384
Quadratic forms,
  372, 382
  a Lemma on, 141
  inequality
    relationship,
    141-143
Quadratic function,
  and conjugate
    vectors, 391-392
Quartic function, 221
Quasi-Newton methods,
  394-407
Regular point, 90, 94
Regularity condition,
  90
Reliability problem,
  293

Reset conditions, 216
Residue matrix, 167
Return function, 4
RMAG, 196
Rosen and Suzuki's
  problem, 234-235
S_1^*, S_2^*, S_3^*, 82
S_a, 79
Saddle point, 62
  of a Lagrangian,
    37-38
  of a modified
    Lagrangian, 38
  of a test problem,
    230
  of an augmented
    Lagrangian, 40, 162
Saturation effect, 13
Scaling, 267, 290
Self-scale variable
  metric algorithms,
    399,
  (see also, SSVM)
Semidefinite, 373
Sensitivity, 110-126
  column sensitivity
    matrix, 113
  macroscopic, 115-117,
    122-123
  management problem,
    118, 122
  microscopic, 111-115
  square sensitivity
    matrix, 113
  via Lagrange
    multipliers, 117-123
Separable function, 15
Separable programming,
  41
Sequential unconstrained
  extremization, 28-37
Sequential unconstrained
  methods, 35, 244, 384
Set,
  notation for, 16

Set-up charge, 13
Shadow prices, 117
Simplex algorithm, 2
Singular matrix,
  372, 389
SLOPE, 196
SSVM algorithm,
  402-403
  performance of,
    408-409
  property 1 (P1) of,
    404
  property 2 (P2) of,
    404
  results, 223-227,
    242-243, 250
Squared-error problem,
  105
Steepest ascent,
  386-387
Stochastic problem,
  267-279, 291
Strict complimentary
  slackness, 150
Structural design,
  261-262
Subroutines,
  AUGLAG, 197-198,
    354-356
  CONGR, 192, 339
  DELTA, 181, 197-198,
    349
  DFPRV, 181, 192-196,
    340-342
  DMAX, 197, 348
  FXNS, 299-300
  GALAG, 181, 198,
    356-358
  GRAD, 300-302
  INITL, 181, 191-192,
    332-338
  interconnection of,
    178
  INVERT, 375
  LPNLP, 180-191,
    327-331

Subroutines,
  notation table,
    187-189
  OUTPUT, 199,
    359-362
  SEARCH, 196,
    343-346
  UPDATE, 198-199,
    350-353
  VALUE, 197, 347
Subset, 16
Sufficient conditions,
  examples of, 97-100
  for inequality
    constraint case,
    69
  for the linear
    problem, 100-102
  general case, 81,
    93-96
  one equality
    constraint case,
    61-62
  second-order, 39
  that x^* be a regular
    point, 90-91
  when not satisfied,
    245-247
SUMT, 244-245
Sylvester's test, 373
Tangent hyperplane, 60
Tangent plane, 91-92
Taylor series, 54-56
  of a dual function,
    158
Test problems,
  constrained, 228-258
  unconstrained,
    219-227
Tolerance assignment,
  110-112, 115-117, 262
Transportation cost
  problem, 46
Transportation networks,
  260

Unconstrained search,
  383-409
Unidirectional search,
  28, 199-214
Unimodal function,
  14, 17
Update criterion,
  in FORTRAN, 307
Variable-metric
  methods, 394-395
  self-scaling,
    399-407
  (see also, SSVM)
Vector,
  column, 369
  nontrivial, 377
  notation, 4
  projection, 378
  row, 369
Vector derivatives,
  379
Vectors,
  conjugate, 390-392
  inner product of,
    377
  orthogonal, 377
  outer product of,
    377
Weibull function, 254
Weights,
  effect on solution,
    230-231
  in FORTRAN, 307-308
  on selection of,
    169-170
  on using finite
    values, 140
  re second-order
    sufficient con-
    ditions, 248
  update procedure,
    150-151, 168-170
Window-scheme, 408
Window test, 207-208,
  213
Wood's function, 220
Worst-case design, 111